AF838582

FAILURE ANALYSIS AND PREVENTION

Edited by **Aidy Ali**

Failure Analysis and Prevention
http://dx.doi.org/10.5772/65149
Edited by Aidy Ali

Contributors

Mohd Nasir Tamin, King Jye Wong, Mahzan Johar, Baris Sayin, Cemil Akcay, Baris Yildizlar, Tarik Serhat Bozkurt, Ivanka Stanimirovic, Zdravko Stanimirovic, Fatemeh Afsharnia, Kary Thanapalan, Mark Bowkett, Kelvii Wei Guo, Rashidi Othman, Mohd Shah Irani Hasni, Ainul Akmar Mokhtar, Muhammad Kamil Kamarul Bahrin, Clementina Igwebike-Ossi, Roselita Fragoudakis, Goran Vukelić, Goran Vizentin

Published by InTech
Janeza Trdine 9, 51000 Rijeka, Croatia

© The Editor(s) and the Author(s) 2017
The moral rights of the editor(s) and the author(s) have been asserted.
All rights to the book as a whole are reserved by InTech. The book as a whole (compilation) cannot be reproduced, distributed or used for commercial or non-commercial purposes without InTech's written permission. Enquiries concerning the use of the book should be directed to InTech's rights and permissions department (permissions@intechopen.com).
Violations are liable to prosecution under the governing Copyright Law.

Individual chapters of this publication are distributed under the terms of the Creative Commons Attribution 3.0 Unported License which permits commercial use, distribution and reproduction of the individual chapters, provided the original author(s) and source publication are appropriately acknowledged. More details and guidelines concerning content reuse and adaptation can be found at http://www.intechopen.com/copyright-policy.html.

Notice

Statements and opinions expressed in the chapters are these of the individual contributors and not necessarily those of the editors or publisher. No responsibility is accepted for the accuracy of information contained in the published chapters. The publisher assumes no responsibility for any damage or injury to persons or property arising out of the use of any materials, instructions, methods or ideas contained in the book.

Publishing Process Manager Romina Skomersic
Technical Editor SPi Global
Cover InTech Design team

First published December, 2017
Printed in Croatia
Legal deposit, Croatia: National and University Library in Zagreb

Additional hard copies can be obtained from orders@intechopen.com

Failure Analysis and Prevention , Edited by Aidy Ali
p. cm.
Print ISBN 978-953-51-3713-9
Online ISBN 978-953-51-3714-6

We are IntechOpen, the world's leading publisher of Open Access books

Built by scientists, for scientists

4,200+
Open access books available

116,000+
International authors and editors

125M+
Downloads

151
Countries delivered to

Our authors are among the
Top 1%
most cited scientists

12.2%
Contributors from top 500 universities

CLARIVATE ANALYTICS
BOOK CITATION INDEX
INDEXED

WEB OF SCIENCE™

Selection of our books indexed in the Book Citation Index in Web of Science™ Core Collection (BKCI)

Interested in publishing with us?
Contact book.department@intechopen.com

Numbers displayed above are based on latest data collected.
For more information visit www.intechopen.com

For EU product safety concerns:
IN TECH d.o.o., Prolaz Marije Krucifikse Kozulić 3, 51000 Rijeka, Croatia,
info@intechopen.com or visit our website at intechopen.com.

Meet the editor

Aidy Ali is a professor of Mechanical Engineering at the National Defence University of Malaysia (NDUM) or known as the Universiti Pertahanan Nasional Malaysia (UPNM). He received his first degree in Mechanical Engineering from the Universiti Putra Malaysia in 1999. He pursued his PhD degree in the year 2003 with his research on "Improving the Fatigue Life of Aircraft Components by Using Surface Engineering" at Sheffield University. He was then appointed as a lecturer at the Universiti Putra Malaysia in 2006, was rapidly promoted to senior lecturer in 2008 and to associate professor in 2010 and was rapidly appointed as a professor in 2012, all within a 6-year period of time. Professor Aidy has more than 17 years of experience in research and teaching in the field of mechanical engineering, especially in mechanical fatigue and fracture of materials. He has published more than 140 journals in Scopus and ISI and 19 books and secured more than 24 research grants. He supervised 16 PhD degree students, 36 master's students and 30 bachelor's degree students. His expertise is related to mechanical materials for defence applications, fatigue, fracture of materials, failure assessment, failure prevention analysis, reliability engineering prediction and crash analysis.

Contents

Preface

Failure analysis has gradually become the crucial method in investigating engineering components or machine failures. The costs of repair and operation breakdowns make an alarming sign for designers, engineers and manufactures on whether or not to keep the existing machines, worthwhile of repairing, or the machines or components need to be redesigned.

This book covers recent advancement methods used in analysing the root of engineering failures and the proactive suggestion for future failure prevention. The techniques used, especially non-destructive testing such X-ray, is well described. The failure analysis covers materials for metal and composites for various applications in mechanical, civil and electrical applications. The modes of failures that are well explained include fracture, fatigue, corrosion and high-temperature failure mechanisms. The administrative part of failures is also presented in the chapter of failure rate analysis.

The book will bring you on a tour on how to apply mechanical, electrical and civil engineering fundamental concepts and to understand the prediction of root cause of failures. The topics explained comprehensively the reliable test that one should perform in order to investigate the cause of machines, component or material failures at the macroscopic and microscopic level.

I hope the material presented in this book, is not too theoretical and you find the case study, the analysis will assist you in tackling your own failure investigation case.

Last but not least, I would like to thank all authors who made their contribution and support for this book to be realised.

Prof. Dr. Aidy Ali
Department of Mechanical Engineering, Faculty of Engineering,
National Defence University of Malaysia (UPNM),
Kuala Lumpur, Malaysia

Chapter 1

Fatigue Failure Analysis of a Centrifugal Pump Shaft

Mohd Nasir Tamin and Mohammad Arif Hamzah

Additional information is available at the end of the chapter

http://dx.doi.org/10.5772/intechopen.70672

Abstract

This chapter deliberates on the systematic processes in failure investigation of engineering components and structures. The procedures are demonstrated in performing failure analysis of a centrifugal pump shaft. The chemical, microstructural, and fractographic analyses provide information on the material science aspects of the failure. The mechanical design analyses establish the cause of failure based on the stress calculations using the strength-of-materials approach. Fatigue analysis using the modified Goodman criterion is employed with consideration of yielding, under the fluctuating load. It is concluded that fatigue crack nucleated in the localized plastic zone at the threaded root region and propagated to cause the premature fatigue failure of the rotor shaft.

Keywords: high-cycle fatigue, mean stress effect, modified Goodman criterion, rotor shaft, stress analysis

1. Introduction

Pumps are commonly used to transport fluids such as water, sewage, petroleum, and petrochemical products. The pumps can be divided into two general categories, namely dynamic pumps and displacement pumps. In a dynamic pump, such as a centrifugal pump, energy is added to the pumping medium continuously and the medium is not contained in a set volume. The energy, in a displacement pump such as a diaphragm pump, is added to the pumping medium periodically while the medium is contained in a set volume. The pump is driven by a prime mover that is either an engine or an electric motor. The capacity of a pump is defined based on the pressure head (in meters) and the maximum delivery flow rate at a specific speed of the shaft. The latter is related to the required power of the prime mover. Typical specifications of some pumps are shown in **Table 1**.

The main components of a centrifugal pump are the rotor assembly and the casing. The pump rotor assembly comprises the shaft, impeller, sleeves, seals, bearings, and coupling halves, as

INTECH
open science | open minds

© 2017 The Author(s). Licensee InTech. This chapter is distributed under the terms of the Creative Commons Attribution License (http://creativecommons.org/licenses/by/3.0), which permits unrestricted use, distribution, and reproduction in any medium, provided the original work is properly cited.

Parameter	*Pump A*	*Pump B*	*Pump C*	*Pump D*
Suction head (m)	33	24	30	32
Delivery volume (m^3/h)	32	120	150	9
Motor power (kW)	7.5	15	37	2.2
Speed (rpm)	2900	1450	1450	2900

Table 1. Typical specifications of centrifugal pump systems for various applications.

illustrated in **Figure 1**. The spiral-shaped casing or volute surrounding the pump impeller serves to collect the fluid discharged by the impeller. The impeller is a rotating set of vanes designed to impart rotation to the mass of the pumping medium. The coupling halves connect the rotor shaft to the output shaft of the motor.

During operation, the engine or the electric motor drives the pump rotor assembly. The rotational kinetic energy is converted to the hydrodynamic energy of the fluid flow. The fluid enters the pump axially through the eye of the casing and is caught in the impeller blades. The fluid gains both velocity and pressure while being accelerated by the impeller. It is then whirled tangentially and radially outward until it leaves through all circumferential parts of the impeller into the diffuser part of the casing. The doughnut-shaped diffuser or scroll section of the casing decelerates the flow and further increases the pressure.

The shaft of the rotor experiences both cyclic flexural load and torsional load during the pumping operation. The start-and-stop cycles could also induce fluctuating stresses with high mean stress level. In this respect, often, the classical high-cycle fatigue analysis is considered in

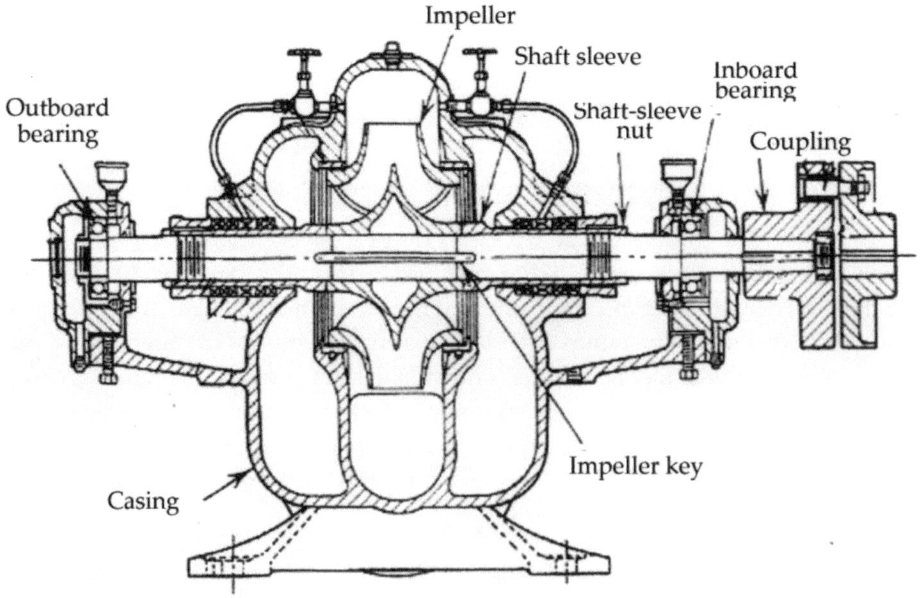

Figure 1. Cut-out section of a centrifugal pump, illustrating the main components.

the safe design of the shaft. The stress analysis on the shaft accounts for the stress raisers due to various design features including fillet, groove, keyway, and screw threads. The presence of such design features, combined with the complex loading, calls for the computational assessment on the reliability of the shaft employing the finite element analysis. In addition, the design should also consider the rotor dynamics aspect (critical speed) of the shaft.

2. Failure and the failure analysis

Experience indicates that despite adhering to the available design procedures, premature failure of the rotor shaft during operation of the centrifugal pump is still reported. The causes of such failure could be classified based on (a) faulty design or misapplication of the materials, (b) faulty processing or fabrication, and (c) the deterioration in service of the component. The severe localized stress due to the design features, as mentioned earlier, could induce excessive plastic deformation leading to the nucleation of fatigue cracks. Fatigue cracks could also nucleate from inherent defects in the material such as nonmetallic inclusions and microvoids, and from machining-induced surface irregularities. In-service deterioration is manifested in the wear of the material, in the form of galling and stress corrosion cracking. The applied fatigue loading continuously causes degradation of the modulus and strength properties of the material. Additional factors that may contribute to early failure of the shaft include poor maintenance of the pump assembly, and improper service and repair of the component.

Procedures	Description	For centrifugal rotor shaft
Description of the failure situation	Background information and service history, records on abnormal operation.	Frequency of start-and-stop operations, details of repair works done, component life.
Visual inspection	Examine the failed component for obvious failure features.	Location of fracture along the shaft, beach lines as sign of fatigue failure.
Mechanical design analysis	Determine if the part is of sufficient size and reliability/life.	Fatigue analysis to demonstrate "infinite life" of the shaft.
Chemical design analysis	Establish the suitability of the material with respect to corrosion resistance.	Determine the chemical composition of the shaft material.
Metallographic examination	To help establish such facts as whether the part has correct heat treatment.	Identify heat treatment of the shaft through the analysis of microstructure and hardness measures.
Determine properties	Determine the properties of the material, pertinent to the design.	Refer to the material data sheets for the grade of the alloy.
Failure simulation	Establish the response of the component under identical loading and boundary conditions.	Finite element simulation of the shaft to establish the stress field in the failed region.
Report writing	Written report detailing the results of the analysis and the causes of failure. May include recommendations to prevent the occurrence of similar failure situation.	To deliberate the rationale on the causes of the premature failure of the shaft.

Table 2. Steps in performing failure analysis of engineering components and structures.

Once the undesirable failure event occurred, assessment on the extent of damage to the component and its impact to the overall integrity of the system is required. Failure analysis on the failed component is performed to determine the root causes of the failure, thus appropriate steps could be implemented to prevent similar occurrence in the future. The process flow of the failure analysis is summarized in **Table 2**.

This chapter describes the procedures and steps in performing failure analysis. A case study on the failure of the rotor shaft of a centrifugal pump is used for illustration. Adequate discussion on the relevant aspects of the analysis in each step is provided. The methodology presented in this chapter could easily be employed and/or extended in performing failure analysis of engineering components and structures.

3. Description of the failure situation

It was reported that the shaft of a centrifugal pump used to pump the blending of hydrocarbons to deliver the final oil product in a refinery has failed during operation. The failure resulted in a fire of the pump and the piping works of the refinery within the unit, with an estimated total loss of USD 48,000. During the last 12 h before the final fracture of the shaft, a total of 12 start-and-stop operations of the centrifugal pump have been scheduled.

The centrifugal pump was installed and commissioned some 30 years ago. There have been three major repairs of the pump involving leaking of the seal. A mechanical seal was installed on the threaded portion of the shaft with a preload corresponding to 25–30% of the material yield strength. However, no reported abnormality on the shaft was recorded.

A typical operation cycle of the centrifugal pump consists of a start-up and running of the pump at a nominal rotor speed of 2975 rpm for 14 h, with a 2-h complete shut-down interval. The pump operates between 5 and 7 days a week throughout the year.

Figure 2. Failure scenario of the centrifugal pump showing the fractured rotor shaft.

The overall view of the failed centrifugal pump is shown in **Figure 2**. The driving side of the fractured rotor shaft has been removed, while the sleeve remained in place.

4. Visual inspection of the fractured shaft

The simplified geometry of the rotor shaft along with major dimensions is illustrated in **Figure 3**. The drive end of the shaft is connected to the shaft of the motor using the coupler. The distance between the bearing supports is 982 mm. The middle section of the stepped shaft with the largest diameter of 65 mm carries the impeller that is positioned in place with a key. The key way has the dimensions of 9 mm radius, length × width of 60 × 18 mm^2, and the depth of 9 mm. Both ends of this section are threaded with M65 × 1.5 threads to receive the mechanical seals (only the critical threaded portion, located on the right side of the section, is drawn). The shaft fractured at the section through the first thread on the drive side as illustrated in **Figure 3**. The fracture plane is oriented with its normal along the longitudinal axis *(z-axis)* of the shaft.

A closer visual inspection of the fractured surface reveals the morphology as shown in **Figure 4(a)**. A greater portion of the surface was flattened and smeared off due possibly to the repeated grinding against the fracture surface of the mating part while the motor runs after the complete

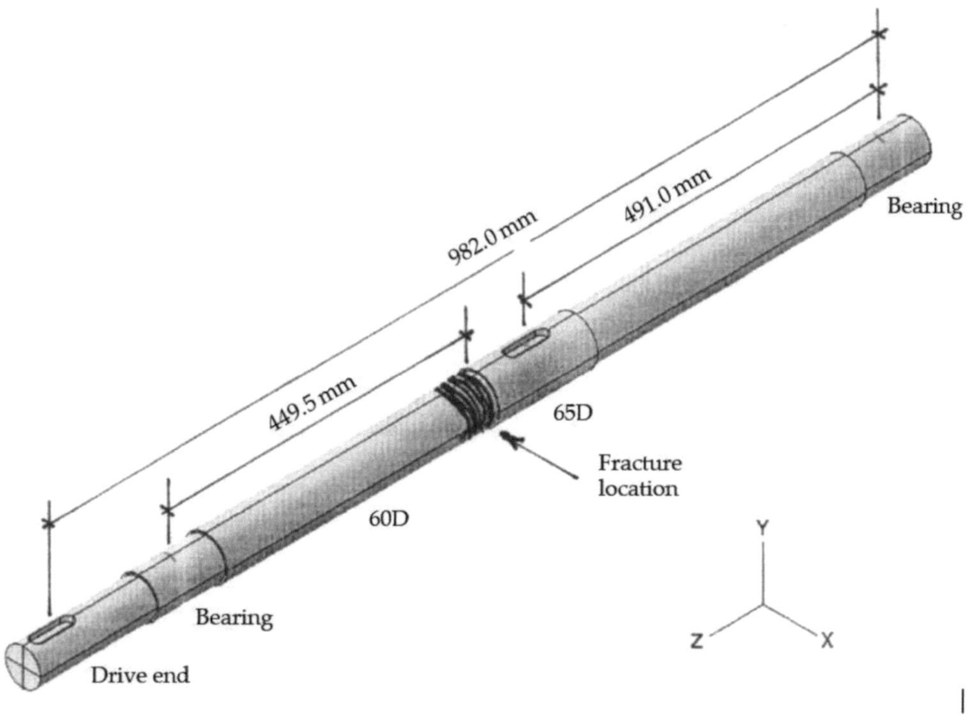

Figure 3. Simplified geometry of the rotor shaft indicating the fracture location.

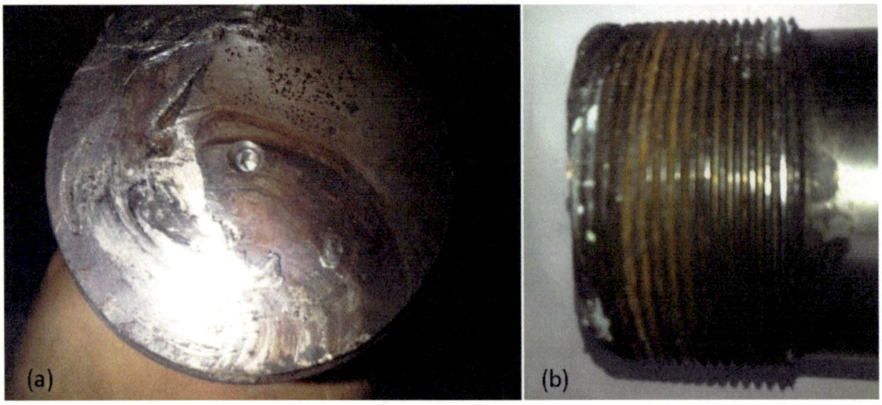

Figure 4. (a) Morphology of the fracture surface, indicating the fatigue *beach lines,* (b) orientation of the fracture plane perpendicular to the longitudinal axis of the shaft.

fracture. Thus, details of the fracture feature could not be extracted easily from the fractograph. However, traces of beach marks indicating fatigue failure are obvious.

It is worth noting that the fracture plane is oriented almost perpendicular to the longitudinal plane of the shaft, as shown in **Figure 4(b)**. Such orientation of the fatigue fracture plane is indicative of the Mode-I (opening) crack propagation under the induced flexural fatigue loading.

5. Chemical design analysis

The chemical design analysis is performed to establish the conformance of the failed shaft material to the manufacturer's materials specification. The manufacturer's record shows that the failed rotor shaft was made of AISI 4140 HT steel. In this respect, the chemical composition of the shaft material is determined using the Glow Discharge Spectrometer (GDS). The resulting chemical composition (in wt. %) is summarized in **Table 3**, the remaining being *Fe*. Other elements detected in the alloy are listed in **Table A1** of the Appendix. The nominal range of the composition for each element of the AISI 4140 steel is also indicated for comparative purposes [1]. It is noted that the failed *Cr-Mo* steel shaft contains slightly higher carbon content, within the range for AISI 4150 steel. Copper is also detected in the failed shaft material.

Microstructures of the steel at two different magnifications are shown in **Figure 5**. The dark and light phases represent tempered and untempered martensite, respectively, with acicular or needlelike structures.

Elem.	*C*	*Mn*	*Cr*	*Mo*	*Si*	*Cu*	*Fe*
Cr-Mo steel shaft	0.502	0.722	1.08	0.228	0.187	0.23	Bal.
AISI 4140	0.38–0.43	0.75–1.00	0.80–1.10	0.15–0.25	0.15–0.30	–	Bal.
AISI 4150	0.48–0.53	0.75–1.00	0.80–1.10	0.15–0.25	0.15–0.30	–	Bal.

Table 3. Elemental composition of the Cr-Mo steel shaft (wt. %) and the reference steel.

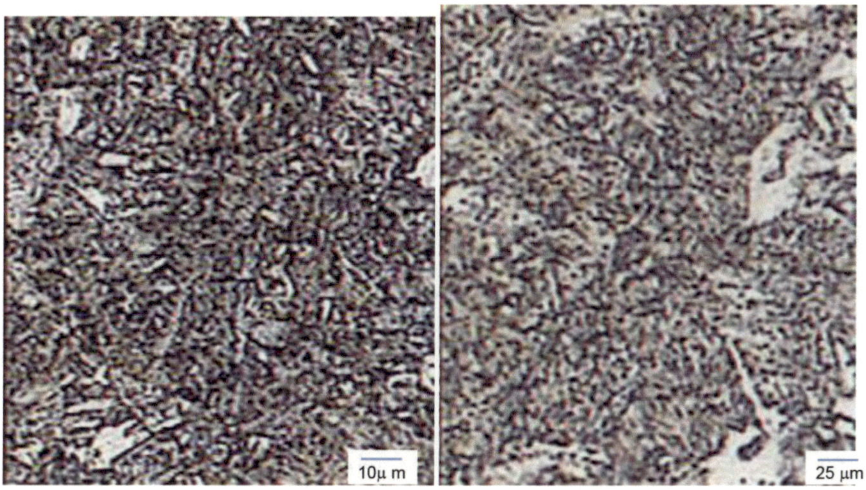

Figure 5. Microstructures of the shaft material showing the matrix of martensite.

Hardness measurements were taken on the failed sample of the steel shaft, at the location near the fractured section. The mean of 20 Vickers hardness readings measured across the section of the shaft is 327.4 HV with a standard deviation of 22.3 HV. The corresponding hardness number on the Brinell and Rockwell C scale is 311 HB and 33 Rc, respectively. Based on the observed microstructures and the hardness measures, it is concluded that the shaft material was likely AISI 4150, oil-quenched, tempered at 595°C. However, the presence of *Cu* could have improved the toughness of the alloy at the expense of lower tensile properties.

6. Mechanical design analysis

Mechanical design analysis is performed to examine the adequacy of the design against yield and fatigue failure of the shaft material. The analyses consist of the stress calculations, particularly at the observed fractured section of the rotor shaft. The critical stress states are then compared to the respective strengths of the shaft material to establish the possible causes of failure. In this respect, the stress levels in the rotor shaft arising from three different load cases are considered, as follows:

i. Stresses during the pumping operation at the rated load. Such stresses could lead to high-cycle fatigue failure of the shaft. The fluctuating load consists of

- a steady torque of T = 834.7 N.m during the power transmission of 260 kW at 2975 rpm from the electric motor to the rotor shaft.
- the cyclic flexural stress induced by the mass of the shaft, assumed to be uniformly distributed along the length of the shaft between the bearing supports (w = 0.12 N/mm). The weight of the impeller generates a concentrated transverse force of 392.4 N.

ii. Stresses during the transient start-and-stop operation: The peak torque at each start-up cycle is up to three times higher in magnitude than that during the nominal operating speed. Such high torque is derived from the inertia effect of the rotor shaft and the massive impeller.

iii. Additional stresses due to preloading of the lock nut for the mechanical seal: The threaded region where fracture was observed was fitted with a locknut that holds the seal in place. The preload contributes a relatively high mean stress to the existing alternating stress component. In addition, the thread geometry inherits the stress concentration at the root region.

The torsional load sequence experienced by the rotor assembly during a typical loading cycle is illustrated in **Figure 6**. It consists of a short start-up process that exerts three times higher peak torsional load than the nominal load level and a 14-h operation at the rated load. This is followed by a complete shut-down for a 2-h interval. Based on the service record that the rotor assembly was recently balanced, thus the contribution to the failure due to the potentially high dynamic imbalance load is ruled out. The rotor assembly has been in service for 30 years and 3 months at the time of fracture. This corresponds to a total of 16,500 start-and-stop cycles endured by the centrifugal pump system.

6.1. Material properties

The required set of mechanical properties of the *Cr-Mo* steel for use in the stress analysis is obtained from published literature [2, 3]. The properties are based on data for AISI 4140, oil-quenched and tempered at 650°C to 285*HB*. The tensile strength (S_U) and yield strength (S_Y) of the material is 758 and 655 MPa, respectively. The cyclic yield strength $\left(S'_Y\right)$ is estimated at 458 MPa.

The endurance limit (S'_e) is reported to be 420 MPa at 10^7 cycles. Since the reported fatigue limit is often established using smooth specimens, it should be corrected to account for the surface condition at the fracture location and the large diameter of the rotor shaft relative to the fatigue test specimens. The surface of the fractured threaded region was machine-finished, and thus the fatigue limit-modifying factor, k_a = 0.72 (refer to **Figure A1a** of the Appendix). Consider the size effect based on the root diameter of the shaft with M65 × 1.5 threads, the corresponding fatigue limit-modifying factor, k_b = 0.795, as determined from **Figure A1b** of the Appendix.

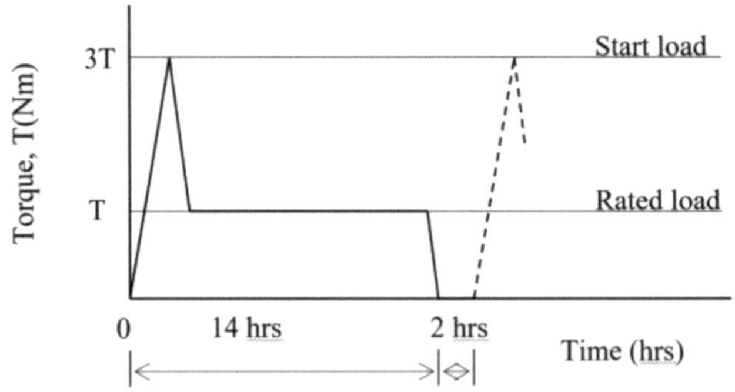

Figure 6. Torsional load sequence experienced by the failed rotor shaft.

The corrected fatigue limit (S_e) is then estimated as

$$S_e = k_a k_b S'_e = (0.72)(0.795)(420) = 240\ \text{MPa} \quad (1)$$

The thread root geometry induces the local stress gradient. Such stress concentration effect is quantified using the fatigue stress concentration factor, K_f. It accounts for the sensitivity of the notched geometry to fatigue stressing. It is defined as the ratio of the endurance limits of notch-free specimens to notched specimens. In this analysis, K_f is treated as a factor that increases the stress, instead of decreasing the fatigue limit of the material. Based on published data for steel-threaded members with hardened and cut threads, the value of K_f is taken as 3.8 (see **Table A2** of the Appendix).

6.2. Fatigue analysis of the shaft at the rated pumping load

Stress analysis is performed for the critical section of the shaft where fracture is observed. Therefore, the calculations are based on the minor diameter of the threaded part for the M65 × 1.5, $d = 63.16$ mm [4]. The bending moment at the critical section is calculated to be $M_C = 102.5$ kN.mm and the corresponding nominal stress is 4.14 MPa. In a rotating shaft, this stress represents the amplitude of the stress cycles. This stress amplitude is further amplified by the fatigue stress concentration ($K_f = 3.8$) as discussed earlier, to give the operating local stress amplitude, $\sigma_a = 15.75$ MPa.

The constant shear stress magnitude of 16.87 MPa is amplified by the geometrical stress concentration associated with the notch root of the thread with $K_{ts} = 2.0$ (see **Figure A2**). This results in the mean shear stress, $\tau_m = 33.74$ MPa.

Since the magnitude of both the normal stress amplitude, σ_a, and the mean shear stress, τ_m, are small relative to the corrected fatigue limit of the shaft material, S_e, the failure of the rotor shaft due to the rated nominal load is ruled out. In fact, this has been demonstrated by the accumulated fatigue life of the rotor at more than 4×10^{10} cycles during the 30 years of service.

6.3. Fatigue analysis due to start-and-stop operations

Duty cycles of the pump consist of the transient start-and-stop operations, as illustrated in **Figure 6**. The start-up procedure exerts three times higher peak torque ($3T = 2504$ N.m) to the shaft due to the inertia effect of the rotor assembly. This corresponds to a maximum shear stress of 101.22 MPa. Thus, the transient start-and-stop cycles induce an alternating shear stress, $\tau_{xy,a} = 50.61$ MPa and a mean shear stress component, $\tau_{xy,m} = 50.61$ MPa. The bending stress amplitude arising from the dead weight of the shaft and impeller remains at $\sigma_a = 15.75$ MPa.

An equivalent fluctuating load, in terms of normal stress components, for the combined loading of fluctuating shear and normal stresses can be defined using the distortion energy theory [5]. The mean, σ'_m, and the amplitude, σ'_a, of the von Mises stress are defined, respectively, as

$$\sigma'_m = \sqrt{\sigma^2_{x,m} + 3\,\tau^2_{xy,m}} = \sqrt{3\,(50.61^2)} = 87.7\text{ MPa} \tag{2a}$$

$$\sigma'_a = \sqrt{\sigma^2_{x,a} + 3\,\tau^2_{xy,a}} = \sqrt{(15.75^2) + 3\,(50.61^2)} = 89.1\text{ MPa} \tag{2b}$$

The possibility of fatigue failure due to the transient start-and-stop cycles could then be examined using the modified Goodman failure criterion. Fatigue failure is likely to occur when the criterion value reaches unity:

$$\frac{\sigma'_a}{S_e} + \frac{\sigma'_m}{S_U} = \frac{89.1}{240} + \frac{87.7}{758} = 0.49 \tag{3}$$

The operating localized stress condition due to the start-and-stop cycles is shown as Point *A* on the fatigue diagram, as illustrated in **Figure 7**.

Since the value of the modified Goodman criterion reaches only 0.49, the accumulated start-and-stop cycle is unlikely to cause the observed fatigue failure. In addition, the rotor assembly has only completed about 16,500 start-and-stop cycles prior to the fracture event.

It is then postulated that the localized stresses at the fractured section must have been higher than that previously calculated. This could have been contributed by the tightening of the mechanical seal unit. Such preloading of the sleeve for the mechanical seal induces a mean normal stress component in addition to the existing alternating stress at the critical thread root location. This postulate is examined in the next section.

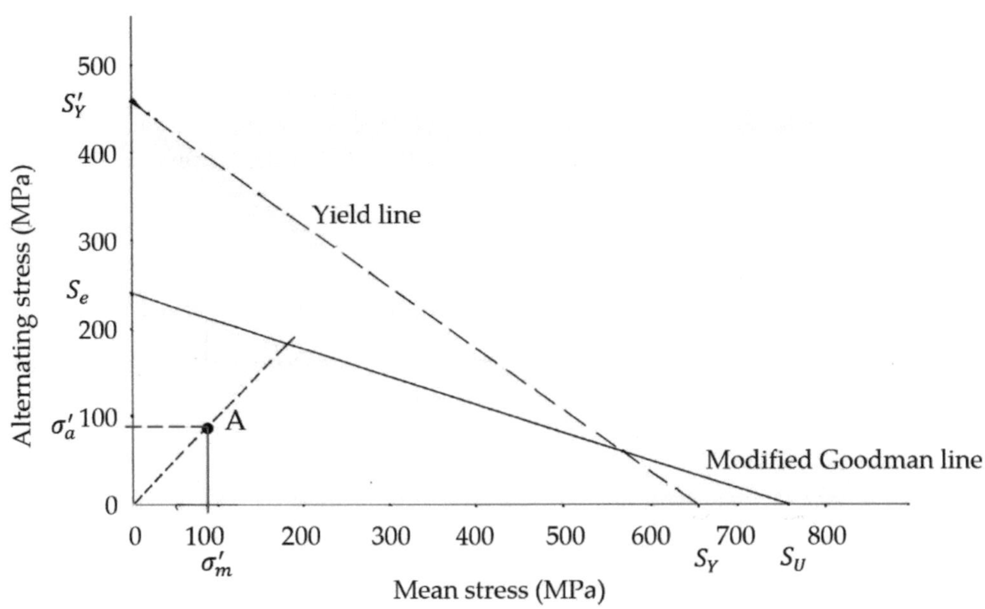

Figure 7. Fatigue life diagram illustrating the modified Goodman failure line and the operating stress condition.

6.4. Contribution of the preload to the local stresses in the threaded region

The tensile force resisted by the threaded shaft, F_b, from the tightening of the sleeve for the mechanical seal can be expressed as the sum of initial tightening force, F_i, and the portion of the externally applied force (bending), F_{ext}, as [6]

$$F_b = F_i + \frac{k_b}{k_b + k_c} F_{ext} \tag{4}$$

The term $\frac{k_b}{k_b+k_c}$ represents the effective stiffness of the threaded shaft and the clamped sleeve of the mechanical seal. The effective stiffness term is proportional to the "effective" shaft section and the clamped area, $C = \frac{A_b}{A_b+A_c}$. The external force represents the induced bending by the dead weight of the rotor assembly. The resisting normal stress in the threaded shaft, $\sigma_b = \frac{F_b}{A_t}$, can then be expressed as

$$\sigma_b = K_t \sigma_i + \frac{A_b}{A_b + A_c} \sigma_{bending} \tag{5}$$

This stress contributes to the mean normal stress component at the critical root region of the threaded shaft. Taking the geometrical stress concentration factor for the cut thread, K_t = 3.0, and estimating the effective threaded shaft section, C = 0.33, the mean stress, $\sigma_{x,m}$, is

$$\sigma_{x,m} = \sigma_b = (3.0)\sigma_i + (0.33)(15.75) \tag{6}$$

The sensitivity of the initial tightening force ($F_i = \sigma_i A_t$) on the resulting fatigue life is examined using Eqs. (2a), (2b), (3), and (6). The allowable initial force is the magnitude to induce the stress up to the proof strength of the shaft material, $S_p = 0.9 S_Y = 589.5$ MPa. The simulated result indicates that an initial preload of the threaded joint to $0.25 S_Y$ would have resulted in an equivalent von Mises fluctuating stresses with $\sigma'_m = 499.1$ MPa, while the alternating component remains at $\sigma'_a = 89.1$ MPa. This stress condition is indicated by Point B as shown in **Figure 8**. The corresponding modified Goodman criterion value of 1.03 suggests a fatigue failure condition at 10^7 start-and-stop cycles. However, we recall that the accumulated start-and-stop cycles at failure are about 16,500 cycles.

The fatigue strength of the shaft material corresponding to the observed finite life of N_f = 16,500 cycles is estimated based on the Basquin equation to be S_{Nf} = 499 MPa. Since this fatigue strength is greater than the cyclic yield stress of the material, the observed failure is likely governed by localized plasticity. This stress condition is represented by Point C in **Figure 8**.

It is concluded that fatigue crack nucleated in the localized plastic zone at the threaded root region of the shaft. The relatively high mean stress component is induced by the preloading of the threaded sleeve. Under continuous fatigue loading, the crack traverses the shaft cross section leading to premature fatigue failure of the rotor shaft.

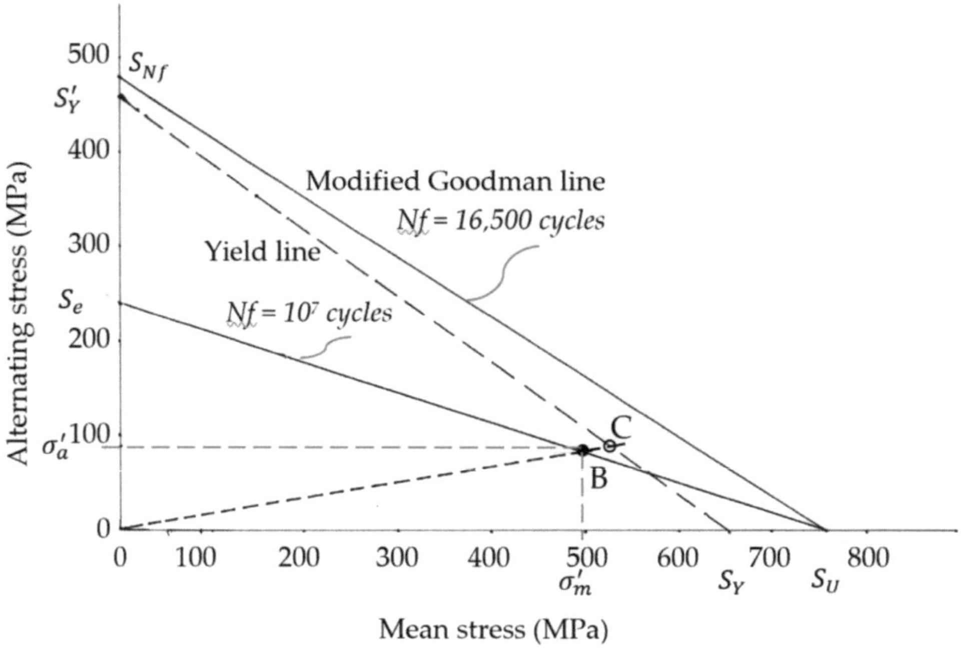

Figure 8. Fatigue life diagram illustrating the possible stress conditions with preload at $0.25S_Y$; fatigue failure (Point B) and yield (Point C).

7. Concluding remarks

Numerous machine components, including the rotor shaft of a centrifugal pump, are load-bearing structures. Since the shaft experiences fluctuating load, fatigue is a common source of the failure. Evidence of fatigue failure is often traceable to the visible beach lines on the fractured surface. However, the cause of failure must be demonstrated through the mechanics-of-deformation analysis involving stress calculations. Unfortunately, the much-needed set of mechanical properties of the material for accurate analysis is often unavailable, thus intelligent estimation is inevitable.

The classical fatigue analysis, as demonstrated in this chapter, is based on constant-amplitude loading. The effect of the transient start-and-stop cycles on the resulting fatigue life of the rotor shaft is not easily accounted for. However, in each analysis involving the rated load cycles and the start-and-stop cycles, premature high-cycle fatigue is an unlikely cause of the failure.

The inherent continuous fatigue degradation of the strength properties is not considered in the strength-of-materials analysis, often performed for failure investigation. However, this classical mechanics analysis is adequate, in most cases, in establishing the cause of failure. The more accurate, yet involved failure prediction based on damage mechanics approach is beyond the scope of this chapter.

http://dx.doi.org/10.5772/intechopen.70672

Appendices

See **Tables A1, A2** and **Figures A1** and **A2**.

	FE (%)	C (%)	Mn (%)	P (%)	S (%)	Si (%)	Cu2 (%)	Ni2 (%)	Cr2 (%)	V (%)	Mo (%)	Ti (%)
Burn 1	96.8	0.509	0.72	0.0137	0.0247	0.188	0.288	0.0619	1.08	0.0079	0.225	0.0116
Burn 2	96.9	0.501	0.723	0.0135	0.0247	0.187	0.205	0.0614	1.08	0.00777	0.226	0.011
Burn 3	96.9	0.497	0.722	0.0134	0.0247	0.186	0.197	0.0614	1.08	0.00758	0.226	0.0109
AVG	96.8	0.502	0.722	0.0135	0.0247	0.187	0.23	0.0616	1.08	0.00775	0.226	0.0112

	Al (%)	Nb (%)	Zr (%)	B2 (%)	B (%)	Sb (%)	Co (%)	Sn (%)	Sn2 (%)	Pb (%)
Burn 1	0.0298	0.00326	0.0112	0.000769	0.000787	0.00696	0.0183	0.0149	0.0115	0.00503
Burn 2	0.03	0.00233	0.0104	0.000723	0.000704	0.00529	0.0182	0.0133	0.0105	0.00418
Burn 3	0.0301	0.00163	0.0097	0.000691	0.000729	0.0053	0.0175	0.0118	0.00949	0.00424
AVG	0.03	0.00241	0.0104	0.000728	0.00074	0.00585	0.018	0.0133	0.0105	0.00449

Table A1. Chemical composition of Cr-Mo steel by GDS analysis.

Hardness	SAE grade (unified thds.)	SAE class (ISO thds.)	K_f rolled thds.	K_f cut thds.
Below 200 Bhn (annealed)	2 and below	5.8 and below	2.2	2.8
Above 200 Bhn (hardened)	4 and above	8.8 and above	3.0	3.8

Table A2. Fatigue stress concentration factors for steel-threaded members [6].

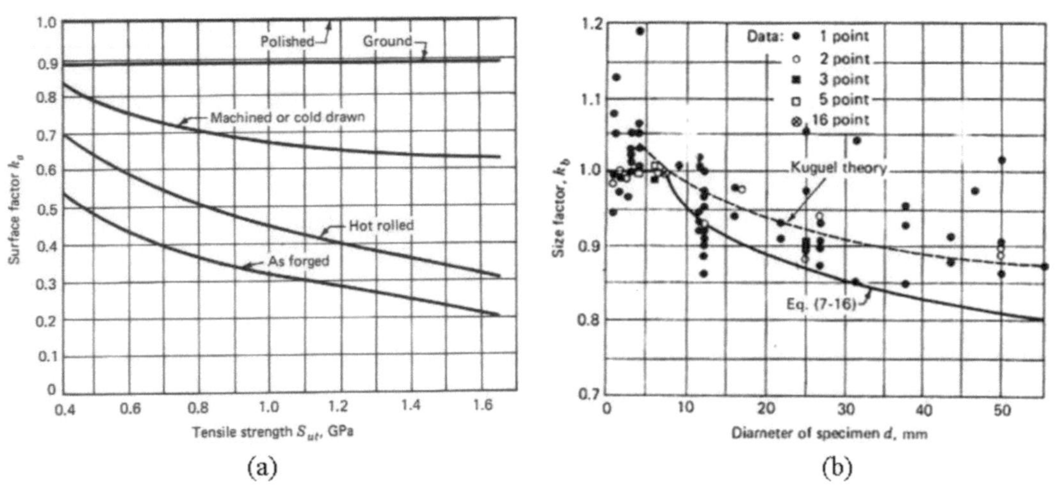

Figure A1. Endurance limit-modifying factors used for the steel shaft [5]. (a) Surface-finish modification factors, (b)Effect of specimen size on the endurance limit in reversed bending and torsion.

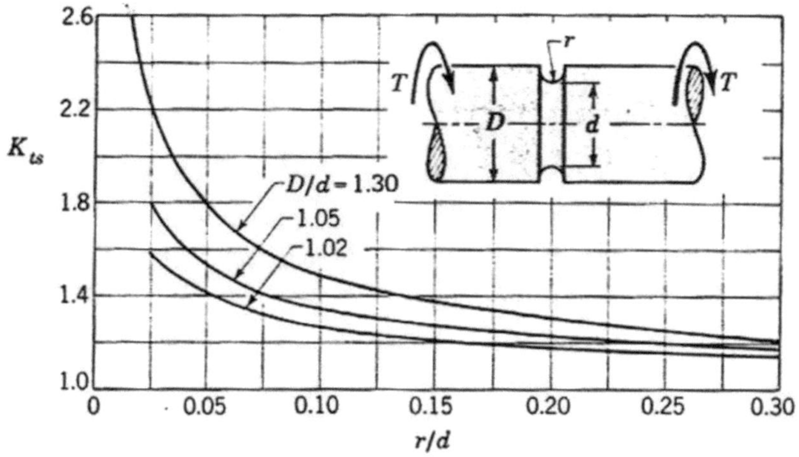

Figure A2. Geometric stress concentration factor chart [5].

Author details

Mohd Nasir Tamin[1]* and Mohammad Arif Hamzah[2]

*Address all correspondence to: nasirtamin@utm.my

1 Faculty of Mechanical Engineering, Universiti Teknologi Malaysia, Johor Bahru, Johor, Malaysia

2 Sarawak Shell Malaysia Berhad, Kuala Lumpur, Malaysia

References

[1] Avner SH. Introduction to Physical Metallurgy. 2nd ed. New York, USA: McGraw-Hill; 1974

[2] Mechanical Properties of 4140 Steel. July 2017. Available from: https://icme.hpc.msstate.edu/mediawiki/index.php/Mechanical_properties_of_4140_steel

[3] Stephen RI et al. Metal Fatigue in Engineering. 2nd ed. New York, USA: Wiley Interscience Publication; 2000

[4] Maryland Metrics Thread Data Charts, Metric Thread-Coarse Pitch-M. July 2017. Available from: https://mdmetric.com/tech/thddat2.htm

[5] Shigley JE. Mechanical Engineering Design. Singapore: McGraw-Hill; 1986

[6] Juvinal RC. Fundamentals of Machine Component Design. New York, USA: John Wiley & Sons; 1983

Chapter 2

Slope Failure Analysis Using Chromaticity Variables

Rashidi Othman and Mohd Shah Irani Hasni

Additional information is available at the end of the chapter

http://dx.doi.org/10.5772/intechopen.71248

Abstract

Slope failure has become a major concern in Malaysia due to the rapid development and urbanisation in the country. It poses severe threats to any highway construction industry, residential areas, natural resources, as well as tourism activities. Thus, this study aims to characterise the relationship between chromaticity variables to be manipulated as indicators to forecast slope failure. The concentration of each soil property in slope soil was evaluated from two different localities that consist of 120 soil samples from stable and unstable slopes located along North South Highway and East West Highway. Indicators that could be used to predict shallow slope failure were high value of variable L*(62), low values of variables c* (20) and h* (66). Furthermore, the hues that indicate stable slope based on Munsell Soil Colour Chart are between 2.5YR and 5YR while the hues that indicate unstable slope are between 5YR and 10YR. The overall analysis leads to the conclusion that the reactions and distinctive changes of chromaticity variables between stable and unstable slopes were emphasised as results of significant differences between soil properties, the locations, slope stability and combinations of all interactions.

Keywords: chromaticity, CIELAB, Munsell Soil Colour Chart, soil properties, shallow slope failure, early warning system, key indicator assessment, Oxisols

1. Introduction

The slope failure trend has increased significantly owing to improper changes in land usage and ranked 10th among the most devastating natural disasters in the world occurring across almost all terrains with steep slopes singled out as the most susceptible to sliding [1]. Marques et al. [2] reported an annual rate of soil erosion of 30–40 ton/ha in developed countries of Asia, Africa and South America. On a global scale, the annual loss of 75 billion tons of soil costs the world about US$400 billion per year or approximately US$70 per person per year [3]. Soil erosion from catchments with natural forests is minimal, but levels of soil erosion tend to increase when natural forest is changed to tree crop plantations.

INTECH
open science | open minds

© 2017 The Author(s). Licensee InTech. This chapter is distributed under the terms of the Creative Commons Attribution License (http://creativecommons.org/licenses/by/3.0), which permits unrestricted use, distribution, and reproduction in any medium, provided the original work is properly cited. (cc) BY

There were many incidents of slope failure occurring both at constructed and natural slopes that caused huge number of deaths especially in tropical countries which received high temperatures and yearly precipitation that brought a large amount of water and consequently triggered extreme effects on the slopes [4]. With these geological factors and climate condition, added with other contributing aspects, slope failure can be considered as one of the crucial threats of environmental catastrophe in Malaysia that requires a serious attention. For example, the slope failures in Hulu Kelang areas have been studied by a number of local researchers and practicing engineers. Ashaari et al. [5] had carried out a field survey work at Hulu Kelang area. A total of 152 slope failures scars of both soil and rock slopes were identified as the potential catastrophe sites. Gue and Cheah [6] investigated the slope failure motions at Kampung Pasir, Hulu Kelang using continuous monitoring approach. They found that the ground had moved from 2 to 17 mm during the monitoring period of 10 days. Hua-xi and Kun-long [7] had performed a detailed investigation on one of the major slope failures occurred in Hulu Kelang area, known as Bukit Antarabangsa 2008 landslide. They concluded that prolonged rainfall during the monsoon season was one of the main factors triggering the failure.

The issue of slope failure in the highway construction industry is closely related to the soil factors [8]. The weakening of soil properties that causes slope failures is resulted from physiochemical activities. It is generated by natural phenomena and human activities through excessive developments which lead to disturbance and destruction of soil surface which are hazardous to slopes. There are many indicators of soil qualities such as organic matters and nutrient deficiency resulted from leachate showing a decline in soil chemical properties while erosion and water infiltration are examples of physical degradation processes [9]. Soil chemical indicators can be identified through specified considerations based on the existence of certain amount of soil colloids whereas physical indicators can be determined by exploring on certain physical appearances and water-holding capacity of the soils. Biological indicators are determined by identifying the amount and mass of microorganisms through concentrations of biogeochemical responses or determining the populations of microorganisms in slope soil.

The soils of the humid tropic such as highly weathered soil (Oxisols) and sandy soil have been observed to be problematic, especially with regard to their fertility. Reviews of research work on current slope soil development in Malaysia, Thailand and Indonesia have significantly shown that such fertility constraints could be improved. Poor fertility of the saprolite is more complex and should be imposed with serious enhancement and management activities. Furthermore, like all acid soils of the humid tropics, Oxisols soils are low in pH value which causes many potential associating problems, including H, Al, and Mn toxicity, Ca deficiency, low CEC, P fixation and low microbial activities [9]. The shallow topsoil is highly vulnerable to erosion and if it is not managed properly especially after the process of clearing the vegetation on top of soil surface, it can slowly lose its original fertility and beneficial physical properties which finally will cause shallow slope failure. Several reviews on the characteristics and management of these soils did not take into account the effect of terracing in exposing saprolites or C horizon. With the surface soils and subsoils already being considered problematic, one could only imagine what kind of impact the saprolites pose to soil fertility.

2. Experimental design

2.1. Description of site selection and soil sampling

Two different localities that were chosen as sampling sites are North South Highway (PLUS) and East Coast Highway (LPT). The whole samples were taken from the slopes that have the gradient lower than 35°. As for unstable soil sample, only slope that collapsed abruptly were collected whereas for stable sample collected from the slope that fully covered by vegetation (**Figure 1**). At North South Highway, 30 soil samples of stable and 30 soil samples of unstable slopes were collected randomly from two sections which were at Section C2 (Tanjung Malim to Bidor) and Section C3 (Sungai Buloh to Tanjung Malim); whereas, in East West Highway (LPT), 30 soil samples of stable and 30 soil samples of unstable slopes were collected from Section 1 (Karak to Jengka) and Section 2 (Jengka to Kuantan). Therefore, a total of 120 soils samplings were collected from those two different localities in Peninsular Malaysia. Auger set was used to collect soil sample at the designated area and soil samples were collected in the depth of 30 cm from the surface. Then, the soil samples were stored in plastic bag and labelled for further analysis.

2.2. Method of data analysis

The collected samples were air-dried, homogenised and sieved to pass a 2 mm mesh sieve for chromaticity variables analysis by CIELAB spectrophotometer. By using a CIELAB spectrophotometer analysis, 10 g of samples were accurately weighed by using analytical balance and was transferred into polystyrene cell and was placed horizontally under spectrophotometer. During the measurement, each sample was measured at three points randomly in order to obtain the mean colour and the variability between different points. When the measurement

Figure 1. Condition and appearance of stable and unstable slopes soil.

was completed, the variables L*a*b*, c* and h* values were displayed on the built by graphical display following each reading. Readings were entered by hand into an Excel spreadsheet.

For analysis, all data gathered were inserted in Microsoft Excel. The mean and standard deviation for each concentration of every experiment was calculated. One-way ANOVA were conducted to measure the validity of the data and the significance of the variation in the results between the stable slope and unstable slope for each soil property.

3. Results

Soil colours give valuable clues in regard to soil properties, soil classification and interpretation. Through this study, the results have been discussed in such a way that it is possible to relate between the Munsell and CIELAB system. It is because, Munsell and CIELAB system has a similar cylindrical structure, and the colour parameters L*, C * [over] ab (CIELAB) as hue, value, chroma (Munsell) represent the same colour perception attributes (hue, lightness, chroma) [10].

3.1. Analysis of soil colour intensity by using Munsell Soil Colour Chart

Scoring with Munsell relies upon human perceived assessment of the three colour attributes: hue, value and chroma. These attributes give valuable clues in soil properties, soil classification and interpretation. Hue is identified as the basic spectral colour or wavelength (Red, Yellow, Blue, or in between, such as Yellow-Red). Value refers to measurement of soil organic matter (OM) in relation to the lightness or darkness of a colour and the range is from 0 (pure black) to 10 (pure white); while chroma is a measurement of colouring agents like Iron or Manganese and the range is from 0 (no colour) to 8 (most coloured).

For this study, the analyses by using Munsell Soil Colour Chart showed a slight difference in stable and unstable slopes. The hues for overall samples were YR (Yellow-Red) and the hues indicating the stable slopes were between 2.5YR and 5YR while the range of hues that indicated the unstable slope was between 5YR and 10YR (**Table 1**). Within each letter range, hue became more Yellow and less Red as the numbers increased. Based on the result, 2.5YR is redder than 5YR and 7.5YR is less yellow than 10YR. This result is consistent with Fontes and Carvallho [11] that reported hue 2.5YR indicates hematite predominate (reddish black), hue 10YR indicates that the soil has goethite (yellowish brown) but does not have hematite whereas hues 7.5YR and 5YR indicate that the soil contains a mixture of goethite and hematite. It is generally believed that hematite, goethite and probably maghemite are the main pigmenting agents in the soil systems [12]. Thus, the different Oxisols variant studied for all sites could be categorised into two main groups which are hematitic or red soil comprising most of the samples from stable slope, and goethite or yellow soils made up most of samples from unstable slope. Therefore, the Munsell chroma combined with the hue value was also used to predict the relative amount of Iron oxides in highly weathered soils [11]. Iron oxides are reddish, yellow and orange in colour [13] and showed in a very small particle size in soils in comparison with other soil minerals which favour their capacity for pigmentation.

No	Munsell soil colour	Hue	Value	Chroma	Colour chips
Stable slopes					
1	Dark reddish grey	2.5YR	3	1	
2	Dusky red	2.5YR	3	2	
3	Dark reddish brown	2.5YR	3	3	
4	Dark red	2.5YR	3	6	
5	Dark reddish grey	2.5YR	4	1	
6	Reddish brown	2.5YR	4	4	
7	Weak red	2.5YR	5	2	
8	Reddish brown	2.5YR	5	3	
9	Reddish brown	2.5YR	5	4	
10	Reddish grey	2.5YR	6	1	
11	Light reddish brown	2.5YR	6	3	
12	Light red	2.5YR	6	6	
13	Very dark grey	5YR	3	1	

No	Munsell soil colour	Hue	Value	Chroma	Colour chips
14	Dark reddish grey	5YR	4	2	
15	Reddish brown	5YR	4	3	
16	Yellowish red	5YR	4	6	
17	Reddish brown	5YR	5	3	
18	Yellowish red	5YR	5	6	
19	Yellowish red	5YR	5	8	
20	Reddish yellow	5YR	7	6	
21	Brown	7.5YR	4	4	
22	Reddish yellow	7.5YR	6	6	
23	Light brown	7.5YR	7	4	
24	Dark brown	10YR	3	3	
25	Dark yellowish brown	10YR	4	4	
26	Brown	10YR	5	3	

http://dx.doi.org/10.5772/intechopen.71248

No	Munsell soil colour	Hue	Value	Chroma	Colour chips
Unstable slopes					
1	Pale brown	2.5YR	8	4	
2	Light reddish brown	2.5YR	7	3	
3	Pinkish white	2.5YR	8	2	
4	Reddish brown	5YR	5	4	
5	Light reddish brown	5YR	6	3	
6	Light reddish brown	5YR	6	4	
7	Pink	5YR	7	4	
8	Reddish yellow	5YR	7	8	
9	Pinkish white	5YR	8	2	
10	Brown	7.5YR	5	3	
11	Light brown	7.5YR	6	4	
12	Reddish yellow	7.5YR 6/6	6	6	

No	Munsell soil colour	Hue	Value	Chroma	Colour chips
13	Pinkish grey	7.5YR	7	2	
14	Reddish yellow	7.5YR	7	6	
15	Reddish yellow	7.5YR	7	8	
16	Yellowish brown	10YR	5	6	
17	Light grey	10YR	7	2	
18	Very pale brown	10YR	8	4	
19	Yellow	10YR	8	6	
20	Yellow	10YR	8	8	

Table 1. Summary of overall soil colour analysis by using Munsell Soil Colour Chart in response to stable and unstable slope conditions.

Ibáñez-Asensio et al. [14] stated that the dark colour of the soil organic matter is caused by the humid acid fraction and a huge amount of calcium carbonate that is also influencing organic matter on lightness. Carbonates of Calcium and Magnesium contribute to the white colour of the soils. Moreover, in terms of the differences in the regression equations, Schulze et al. [15] pointed out that the relationship between the organic matter content and the Munsell value of soils was strongly influenced by soil texture, parent material and vegetation. High contents of clay and sand affects the soil colour to become yellowish, reddish and whitish. Clay is the smallest particle in soils and exhibits colloidal properties. Some of the clays, like Iron oxide clay, play an important role in soil aggregation and in addition impart red to yellow colours to soils. Embrapa [16] stated that most minerals are not highly coloured and when they are coated with humus and Iron oxides, they take on the colours of humus (black or brown), Iron oxides and hydroxides (red and yellow). **Table 1** showing summary of overall colour analysis by using Munsell Soil Colour Chart for 120 soil samples in response to stable and unstable slopes.

3.2. Analysis of reflectance colorimeter measurement by using CIELAB spectrophotometer

3.2.1. Analysis of CIELAB L (lightness) value*

Statistical analysis showed that there was a significant difference for Oxisols colour variable L* between stable and unstable slopes. The value of colour variable L* was the lowest in stable slopes for each study area compared to unstable slopes. The value of variables L* in stable slopes ranged from 44 to 50 whereas in unstable slopes, from 61 to 63 (**Figure 2**).

3.2.2. Analysis of CIELAB a (red-green axis) value*

Statistical analysis showed that there was a significant difference for Oxisols colour variable a* between stable and unstable slopes. The values of colour variable a* were higher in stable slopes from sites, Plus C2 and Site 2 LPT in comparison to unstable slopes. However, the results also revealed that the values of variable a* in stable slopes from sites, Plus C3 and Site 1 LPT were lower compared to unstable slope respectively. The level of variable a* in stable slopes ranged from 7.2 to 11.8 whereas in unstable slopes, from 5.4 to 10.5 (**Figure 3**).

3.2.3. Analysis of CIELAB b (yellow-blue axis) value*

Statistical analysis showed that there was a significant difference for Oxisols colour variable b* between stable and unstable slopes. The values of colour variables b* were higher in unstable slopes for each study area compared to stable slopes. The values of variable b* in unstable slopes ranged from 19 to 22.7 whereas in stable slopes, from 16.5 to 19.8 (**Figure 4**).

SITE	PLUS C2	PLUS C3	SITE 1 LPT	SITE 2 LPT
STABLE SLOPE	49	50	44	48
UNSTABLE SLOPE	63	61	62	63

Figure 2. The average of variable L* in Oxisols for stable and unstable slopes.

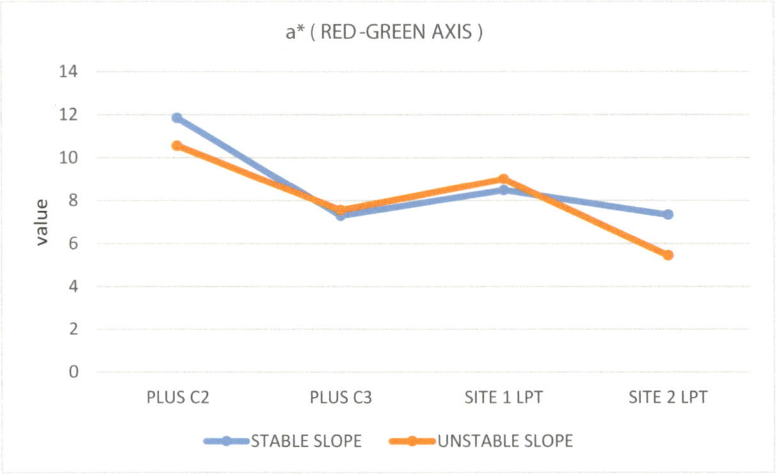

SITE	PLUS C2	PLUS C3	SITE 1 LPT	SITE 2 LPT
STABLE SLOPE	11.8	7.2	8.5	7.3
UNSTABLE SLOPE	10.5	7.5	8.9	5.4

Figure 3. The average of variable a* in Oxisols for stable and unstable slopes.

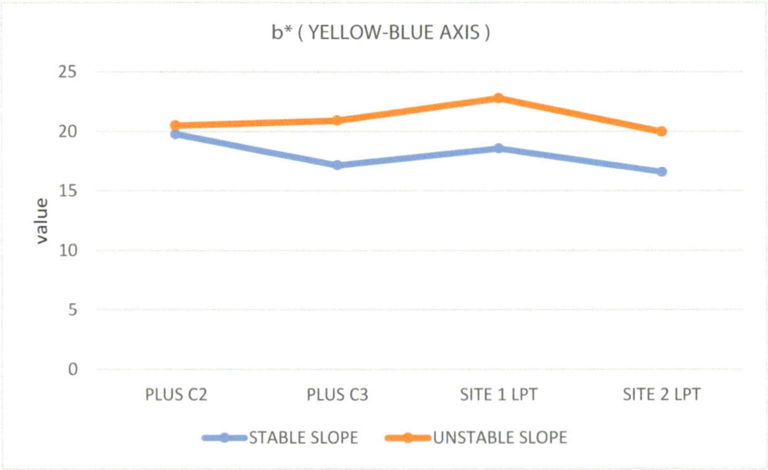

SITE	PLUS C2	PLUS C3	SITE 1 LPT	SITE 2 LPT
STABLE SLOPE	19.8	17	18.5	16.5
UNSTABLE SLOPE	20.4	20.9	22.7	19

Figure 4. The average of variable b* in Oxisols for stable and unstable slopes.

3.2.4. Analysis of CIELAB c (chroma) value*

Statistical analysis showed that there was a significant difference Oxisols colour variable c* between stable and unstable slopes. The values of colour variable c* were higher in stable slopes for each study area compared to unstable slopes. The values of variable c* in stable slopes ranged from 21 to 25 whereas in unstable slopes, from 18 to 23 (**Figure 5**).

3.2.5. Analysis of CIELAB h (hue) value*

Statistical analysis showed that there was a significant difference for Oxisols colour variable h* between stable and unstable slopes. The values of colour variable h* were higher in stable slopes for each study area in comparison to unstable slopes. The values of variable h* in stable slopes ranged from 65 to 76 whereas in unstable slopes, from 60 to 69 (**Figure 6**).

3.2.6. Overall CIELAB soil colour value analysis

Statistical analysis showed that there was a highly significant difference ($P < 0.0001$) in overall CIELAB soil colour variables analysis between stable and unstable slopes. The results of the quantitative measurements of soil colour performed in the laboratory are summarised in **Table 2**. The value of variables L* was higher in unstable slopes rather than stable slopes with the values of 62 and 48, whereas the value of variables a* was slightly lower in unstable slope with the value of 8.1 compared to the stable slope with 8.7 in value. The value of variables b* was slightly higher in unstable slope in comparison to stable slope with the value of 21 and 18. Finally, the value of variables c* and h* were lower in unstable slopes as compared to the stable slopes. In conclusion, the unstable slopes for overall sites consist of higher value of colour variables L*, b* and lower value of colour variables a*, c* and h*, respectively. Since there is a significant difference of colour variables in the comparison of different slope condition, it is possible to conclude the soil colour can be an indicator for early warning of shallow slope failure.

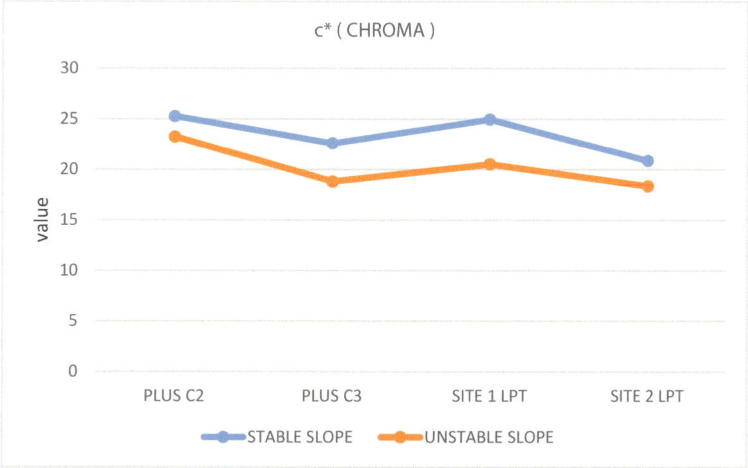

SITE	PLUS C2	PLUS C3	SITE 1 LPT	SITE 2 LPT
STABLE SLOPE	25	23	25	21
UNSTABLE SLOPE	23	19	21	18

Figure 5. The average of variable c* in Oxisols for stable and unstable slopes.

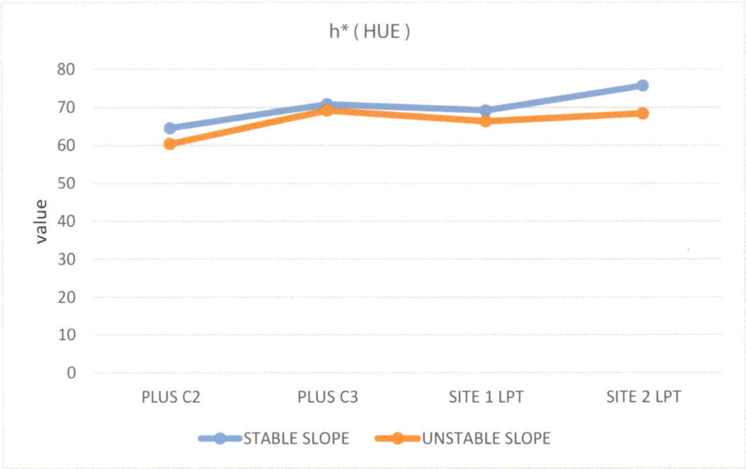

SITE	PLUS C2	PLUS C3	SITE 1 LPT	SITE 2 LPT
STABLE SLOPE	65	71	69	76
UNSTABLE SLOPE	60	69	66	68

Figure 6. The average of variable h* in Oxisols for stable and unstable slopes.

Colour variables	Slope condition	
	Stable slope	Unstable slope
L* (lightness)	48 ± 6	62 ± 6
a* (red-green)	8.7 ± 2	8.1 ± 1.5
b* (yellow - blue)	18 ± 4	21 ± 4
C* (Chroma)	23 ± 4	20 ± 4
h* (hue)	70 ± 6	66 ± 6

Table 2. Mean values for overall soil colour variables in stable and unstable slopes.

4. Discussion

4.1. Relationships between CIELAB spectrometer and Munsell Soil Colour Chart

Munsell Soil Colour Charts are develop for colour identification of an object by direct comparison by using set of colour palettes with a sequence of colour samples on each page in it. CIELAB, is the methods that operated without relying on human eye. It works by scanning an object via spectrophotometer and the outcomes were recorded in graph in three dimensional colour space. This equipment is very effective as supporter to the Munsell colour system. Some complications that make the outcomes an alternative and attractive prospect are inherent to the Munsell colour system for example, a great degree of subjectivity and unpredictability between researchers. CIELAB spectrophotometer able to capture more colour data than Munsell colour charts because the level of precision by CIELAB spectrophotometer is available in colour description.

Through the result and analysis using CIELAB spectrometer, variation of soil colours became apparent at different slope conditions. The soil samples collected from the unstable slopes were characterised by the high values of variables L*, b* and slightly lower values of variables a*, C* and h* (**Table 2**). The difference was also detectable by the soil colour reader using the Munsell colour system, where the findings indicated that the hues for overall samples were classified as YR (Yellow-Red), with the stable 2.5YR to 5YR whereas unstable slope 5YR to 10YR, respectively. The positive value CIELAB variables a* and b* are the indication that this soil sample of Oxisols are dominance by the Iron oxide which influencing the soil colour and also can detected directly through the Munsell Soil Colour Chart. This difference can be measured through the b* value where a strong positive value indicates a strong yellow colour and a strong negative value indicates a strong blue colour whereas a* value where a positive value indicates a red colour and negative value indicates green colour [17].

Moreover, the high value of variables L* in unstable slope in comparison to stable slope is the indication that the unstable slope consist low amount of organic matter content which have a great influencing as among the colouring agent for Oxisols soil. These observations also suggest that the colour of the slope soil samples, particularly the variable L* as soil colour that attributes lightness (similar to Munsell value) provides the most information about the relationship with soil properties which accounted to the ranges of 61–63. Since there was a significant difference for colour variables in the comparison of different slope conditions, it is possible to conclude the soil colour can be an indicator for early warning of shallow slope failure.

4.2. Relationship between chromaticity variables and soil properties

The study on the relationship between chromaticity variables and other soil properties was initiated by investigating the response of soil colour, in stable and unstable slopes samples, in relation to some general properties of the slope soils. CIELAB analysis and Munsell Soil Colour Chart tests revealed important relationships between chromaticity variables and soil properties. These observations suggest that the variable L* provides marked indicator about the relationship with most of the soil properties. Colour variable L* is closely correlated with soil texture including clay and sand contents, soil organic carbon, Iron oxide and Aluminium concentrations. The relationships between L*, Iron oxide and Aluminium which contributed to the darkening of the soils, either individually or associated with other organic materials; also had been revealed by Ibáñez-Asensio et al. [14]. Furthermore, colour variable L* was found strongly affected by differences in climate and vegetation as well as soil moisture regime. The L* variable was the lowest (darker) in stable slope soils which covered by vegetation particularly with fern species in comparison with those unstable slope soils without vegetation. This attributed to the effect on the soil colour of the variations in the composition and quantity of soil humus [17].

Therefore, based on the results, it can be pointed out that unstable slope higher value of variable L* (lightness), which was the evidence that soil properties have strong influence on the weakly developed soils. Several researchers found the lightness value decreased when the number of clay sized particles increased and also indicated the presence of Iron oxides [15]. Most of Iron oxides are seen in small particle size so that very small quantities would be adequate in influencing soil colour since some of those particles had remained on the negatively charged surface of clay grains [15]. The correlation of L* with clay content, may be caused by the relationship of free oxides and humid compounds with phyllosilicates, favoured by their superficial activities. The phyllosilicates act as a support for the other pigments as the increase in phyllosilicates content (white or greyish pigments) would also cause an increase in L* [17].

Iron oxide and soil total organic carbon have a combined influence on L*, and this effect is closely linked to the soil particle sizes. Schulze et al. [15] had found that soil organic carbon and Iron influence on the spectral reaction of soils interrelates with particle size. The results of the present research are consistent with their findings as it is confirmed that interactions between texture and Iron exist. It also showed a positive relationship between soil lightness and soil texture. Sawada et al. [17] also had found low values for soil colour variables particularly L* in stable slope soils with high clay content, high organic carbon and high concentration of Iron oxide. On the contrary, unstable slope contains high values of colour variables, high sand contents, low organic carbon and less concentration of iron oxide. Soil with coarse textures and low levels of organic carbon, consequently, generated greater lightness values than in soils with finer textures and higher contents of organic matter. The analysis used in this study confirms the multivariate relationships and the outcomes pointing to this direction. Finally, in relation to soil organic carbon content effects on lightness, this research had discovered the same relationship found by Konen et al. [18].

Moreover, in this study, it was found that variable h* (Hue) of the stable slope was slightly higher in comparison to unstable slope. This could be caused by the joint migration of the clay and Iron formed in the slope soils. The higher duration of the annual dry period at altitudes of less than 1000 m, higher temperatures with the consequent rising in dehydration of the Iron forms, could be the factors of this greater reddening [13]. This result is also consistent with the finding through the Munsell Soil Colour Chart that indicates the hues for overall samples were YR (Yellow-Red) which were influenced by high concentration of Iron oxides in studied area. Iron oxides can reflect the surrounding of the environments in which they are formed and are considered as colouring agents for most of slope soil samples. Curi [12] stated that in the soil systems, hematite, goethite and probably maghemite which are classified under Iron oxides are the main pigmenting agents in influencing soil colour. The hues that indicate stable slope are between 2.5YR and 5YR and unstable slope are between 5YR and 10YR. 10YR hue shows that the soil contains goethite predominance while 2.5YR hue is hematite predominate, whereas hues 7.5YR, and 5YR show a combination of goethite and hematite. From the findings it can be concluded that the yellowish colour for most of the unstable slope soil samples are caused by a yellow to brown iron oxide mineral called goethite. Generally, these soils have lower iron contents extracted by the sulphuric acid digestion than the other. That occurs either because the parent material had a low iron content or because iron was removed from the soil by percolating water.

Due to the yellower colour, it is relatively easy to distinguish the horizons for instance the red colour dominance for most of the stable slope soil samples are due to hematite and a dark red due to iron oxide. The content of iron oxides extracted by Non-ferric Red Oxisols are quite variable in texture, which ranges from medium to very clayey. The parent material for these soils is very variable and ranges from sandstones to pelitic rocks, with the major requirement being relatively high iron content. Similar to the variable a* value in stable slopes which is slightly higher and this means that the soils are slightly redder in comparison to unstable slopes which contain high sand fraction and slightly higher b* values, it means that the soil is dominantly yellowish in colour [18].

In addition, with respect to the overall soil samples, the conditions of slope are essential in influencing the chroma values in such a way that the stable slopes has a slightly high chromatic value than the unstable slopes. As was stated before, this is interrelated to the larger amount of Iron formed in the stable slopes which might be caused by modification and illuviation of slope structure. Chroma C* is also correlated with total organic carbon content, which has also been described for other soils [19] and most strongly associated with soil texture content. This means that the colour of the texture becomes more uniform as the contents of these minerals increases, which may be attributable to the processes of reduction and enlargement.

5. Conclusion

The overall soil colour analyses lead to the findings that the CIELAB variables L* (lightness), c* (chroma) and h* (hue) with significant values of colour variables measured at different slope conditions, provide the most information in relation to soil properties. Regarding to the relationship between soil properties, the study had identified that soil texture, total organic carbon (TOC), Iron oxide and Aluminium concentration were strongly interrelated with soil colour variables at the studied areas. It is also recommended that in order to explain and detect the colour of slope soils, the function and availability of lightness-darkness as an analytical factor should be highlighted, together with the amount of Iron oxides. To sum up, indicators that can be used to predict shallow slope failure based on chromaticity variables are high value of variable L*, low values of variables c* and h* and the Munsell Soil Colour Chart were between 2.5YR and 5YR (hematite predominance). The correlative relationships between chromaticity variables and soil erosion suggest that all these properties may potentially be used as an indicator of slope failure.

These findings should be used to trigger further investigation of the reasons or sources for the failure of the slope soil and an assessment made of the potential risks to humans or the environment if the failure continues. Through this study it showed that the weakening of the slope soil properties occurred mostly due to erosion effect towards the existing soil properties. Consequently, serious attention should be emphasised on each slope along the highways particularly the unstable slopes in order to reduce harmful effects. Most of the landslides occurred during the rainy days when the soil is relatively wet. It would require special preventing strategies such as slope levelling, terracing and practicing in planting suitable vegetation in slope areas.

Vegetation and slope stability are interrelated by the ability of the plant life growing on slopes to both promote and hinder the stability of the slope. The relationship is a complex combination of the type of soil, the rainfall regime, the plant species present, the slope aspect, and the steepness of the slope. Any study of soil properties should take serious attention towards any vegetation above the slope area as this factor is crucial in influencing the loss of several nutrients. Planting vegetation will increase the organic carbon in soil thus the ions of organic carbon will bind with ion in clay and hydrogen in soil. These reactions will strengthen the soil structure. Knowledge of the underlying slope stability as a function of the soil type, its age, horizon development, compaction, and other impacts are the major underlying aspect of understanding how vegetation can alter the stability of the slope. Our study did take note of vegetation, but for future studies, a more thorough study with regard to the vegetation of the areas in conjunction with certain soil properties would be interesting to be highlighted. The research findings showed that unstable slope was more likely to occur if there is no plant life growing on the top of soil. The less vegetation growing in the soil the more likely that erosion will happened. Vegetation can protect the soil from the impact of the rain and slows down the infiltration process. Plants with deeper roots are better at holding the soil together and protect it from erosion.

Finally, an efficient management on landslide risk, the coordination between regions, departments concerned, universities, research centres, non-governmental organisations and local peoples in landslide-prone would be helpful in order to obtain the better risk management. This coordination and communication would minimise the wasting budget, man power, time allocation and miscommunication of decisions taken in future. Additionally, the findings of this research can be integrated with the various components of landslide risk in risk information and management systems which should be developed as spatial decision support systems for local authorities dealing with risk management. The availability and quality of historic landslide database cannot be overemphasised since they constitute the basis for all components of landslide. Modern technologies, such as geographic information system (GIS) and remote communications, should have a wider application in landslide risk assessment and management.

Acknowledgements

The authors would like to thanks Ministry of Education (MOE) and International Islamic University Malaysia (IIUM) for the Research Grant NRGS13-002-0002.

Author details

Rashidi Othman* and Mohd Shah Irani Hasni

*Address all correspondence to: rashidi@iium.edu.my

International Institute for Halal Research and Training (INHART), Herbarium Unit, Department of Landscape Architecture, Kulliyyah of Architecture and Environmental Design, International Islamic University Malaysia, Kuala Lumpur, Malaysia

References

[1] Leroy SA, Gracheva R. Historical events. In: Encyclopedia of Natural Hazards. Netherlands: Springer; 2013. p. 452-471

[2] Marques AMJ, Bienes R, Jimenez L, Perez-Rodríguez R. Effect of vegetal cover on runoff and soil erosion under light intensity events. Rainfall simulation over USLE plots. Science of the Total Environment. 25 May 2007;**378**(1–2):161-165

[3] Eswaran H, Lal R, Reich PF. Land degradation: An overview. In: Bridges EM, Hannam ID, Oldeman LR, Pening de Vries FWT, Scherr SJ, Sompatpanit S, editors. Responses to Land Degradation. Proc. 2nd. International Conference on Land Degradation and Desertification; Khon Kaen, Thailand. New Delhi, India: Oxford Press; 2008

[4] Jordán A, Martínez-Zavala L. Soil loss and runoff rates on unpaved forest roads in southern Spain after simulated rainfall. Forest Ecology and Management. 2008;**255**(3–4):913-919

[5] Ashaari M, Shabri S, Mahadzer M, Nik Ramalan NH, Fadlee MB, Mariappan S, Low TH, Chong S, Zaini MM, Wan Mohd RWI. Slope field mapping and finding at Ulu Klang area, Malaysia. In: Proceeding of the International Conference on Slopes Malaysia; Kuala Lumpur, Malaysia. 4–6 November 2008

[6] Gue SS, Cheah SW. Geotechnical challenges in slope engineering of infrastructures. In: International Conference on Infrastructure Development. Malaysia: Putrajaya Mariott Hotel. 2008; 7–9 May 2008

[7] Hua-xi G, Kun-long Y. Study on spatial prediction and time forecast of landslide. Natural Hazards. 2014;**70**:1-14

[8] Harwant S. Slope Assessment Systems: A Review and Evaluation of Current Techniques Used for Cut Slopes in the Mountainous Terrain of West Malaysia. Faculty of Resource Science and Technology. Kuching, Sarawak, Malaysia: University Malaysia Sarawak; 2006

[9] Shakilah N. Studies on chemical properties of stable and unstable slope of highly weathered soil (oxisols) [Dissertation]. Kuala Lumpur, Malaysia: International Islamic University Malaysia; 2014

[10] Sanchez-Mara NM, Delgado G, Melgosa M, Hita E, Delgado R. Cielab colour parameters and their relationship to soil characteristics in mediterranean red soils. Soil Science. 1997;**162**:833-842

[11] Fontes MPF, Carvallho AI Jr. Colour attributes and mineralogy characteristics, evaluated by radiometry of highly weathered tropical soils. Soil Science Society of America Journal. 2005;**69**:1162-1172

[12] Curi N. Lithosequence and toposequence of oxisols from Goiás and Minas Gerais States, Brazil [Unpublished PhD's dissertation]. Indianapolis, USA: Purdue University; 1983

[13] Schwertmann U. Relations between iron oxides, soil colour and soil formation. In: Bigham JM, Ciolkosz EJ, editors. Soil Color. Madison: Soil Society of America; 1993. p. 51-70. SSSA Special Publication No 31

[14] Ibáñez-Asensio S, Marqués-Mateu A, Moreno-Ramón H, Balasch S. Statistical relationships between soil colour and soil attributes in semiarid soils. Biosystems Engineering. 2013;**116**:120-129

[15] Schulze DG, Nagel JL, Van Scoyoc GE, Henderson TL, Baumgardner MF. Significance of organic matter in determining soil colors. In: Bigham JM, Ciolkosz EJ, editors. Soil Color. Madison: Soil Society of America; 1993. p. 71-90. SSSA Special Publication No 31

[16] Embrapa. Manual de métodos de análise de solos. Rio de Janeiro: Centro Nacional de Pesquisas de Solos; 1997. p. 212

[17] Sawada K, Wakimoto T, Hata N, Taguchi S, Tanaka S, Tafu M, Kuramitz H. The evaluation of forest fire severity and effect on soil organic matter based on the L*, a*, b* color reading system. Analytical Methods. 2013;**5**:2660-2665

[18] Konen ME, Burras CL, Sandor JA. Organic carbon, texture, and quantitative color measurement relationships for cultivated soils in North Central Iowa. Soil Science Society of America Journal. 2003;**67**(6):1823-1830

[19] Dobos RR, Ciolkosz EJ, Waltman WJ. The effect of organic carbon, temperature, time, and redox conditions on soil color. Soil Science. 1990;**150**:506-512

Mixed-Mode Delamination Failures of Quasi-Isotropic Quasi-Homogeneous Carbon/Epoxy Laminated Composite

Mahzan Johar, King Jye Wong and
Mohd Nasir Tamin

Additional information is available at the end of the chapter

http://dx.doi.org/10.5772/intechopen.69440

Abstract

This chapter characterised the delamination behaviour of a quasi-isotropic quasi-homogeneous (QIQH) multidirectional carbon/epoxy-laminated composite. The delaminated surface constituted of 45°//0 layers. Specimens were tested using mode I double cantilever beam (DCB), mode II end-notched flexure (ENF) and mixed-mode I+II mixed-mode flexure (MMF) tests at constant crosshead speed of 1 mm/min. Results showed that the fracture toughness increased with the mode II component. Specifically, the mode I, mode II and mixed-mode I+II fracture toughness were 508.17, 1676.26 and 927.52 N/m, respectively. When the fracture toughness values were fitted using the Benzeggagh-Kenane (BK) criterion, it was found that the best-fit material parameter, η, was attained at 1.21. Furthermore, fibre bridging was observed in DCB specimens, where the steady-state fracture toughness was approximately 80% higher compared to the mode I fracture toughness. Finally, through scanning electron micrographs, it was found that there was resin-rich region at the crack tip of the specimens. In addition, fibre debonding of the 45°layer was found to be dominant in the DCB specimens. Significant shear cusps were noticed in the ENF specimens. As for the MMF specimens, matrix cracking and fibre debonding of the 0°layer were observed to be the major failure mechanisms.

Keywords: mixed-mode delamination, quasi-isotropic quasi-homogeneous, laminated composite, BK criterion, scanning electron micrographs

1. Introduction

Carbon fibre reinforced polymer (CFRP) composites are widely employed in advanced structural applications such as aircraft wing skin and fuselage, automobile body panels and marine deck structures. Composites have the advantages of high specific strength and stiffness over

INTECH
open science | open minds

© 2017 The Author(s). Licensee InTech. This chapter is distributed under the terms of the Creative Commons Attribution License (http://creativecommons.org/licenses/by/3.0), which permits unrestricted use, distribution, and reproduction in any medium, provided the original work is properly cited.

conventional aluminium alloy counterparts. This could greatly reduce the weight of the structure, which in turn benefits the industry through cost saving. For example, it was reported that the weight reduction of 1 kg in an aircraft structure can save over 2900 l of fuel per year [1, 2]. However, delamination is generally recognised as one of the earliest failure modes in laminated composites. This is caused by relatively low interlaminar strengths [3]. Delamination could be induced by manufacturing defects or low velocity impacts [3, 4]. This kind of damage is not easily detected [3]. Hence, the separation of the two neighbouring layers has been described as the most feared failure mechanism in laminated composites [5]. Therefore, it is essential to understand the delamination behaviour in laminated composites.

In addition, in the use of laminated composites, multidirectional laminates have been generally known to have higher interlaminar fracture toughness compared to the unidirectional laminates [6, 7]. This could be attributed to relatively blunt crack tip or intralaminar crack [7]. To study the delamination behaviour of multidirectional laminates, characterisation was commonly conducted on pure mode I, pure mode II and mixed-mode I+II tests [4, 8, 9]. It has been mentioned that the mode III component in aircraft structures could be negligible [8]. Therefore, the above-mentioned three delamination modes are the fundamental and essential loading modes to be considered.

In this chapter, the delamination behaviour of a quasi-isotropic quasi-homogeneous (QIQH) laminated composite with fibre orientation of 45° and 0° adjacent to the pre-crack was investigated. Mode I, mode II and mixed-mode I+II delamination behaviour were characterised by double cantilever beam (DCB), end-notched flexure (ENF) and mixed-mode flexure (MMF) tests. Subsequently, the variation of fracture toughness with respect to the mixed-mode ratio was fitted using the Benzeggagh-Kenane (BK) criterion. It was followed by the characterisation of R-curve behaviour of the composite. Finally, fractographic analyses were carried out on the delaminated surfaces.

2. Materials

The composite material used in this study was T600S/R368-1 carbon/epoxy prepreg supplied by Structil. The density of the prepreg was 170 g/m^2, the glass transition temperature of the resin was 105°C, the fibre volume fraction was 59 $\pm$ 2% and the average ply thickness was 0.2 mm. A 48-ply unidirectional composite laminate was fabricated using a hand lay-up technique with the stacking sequence of [0/45/90/-45/90/-45/45/-45/0/90/0/45/0/45/-45/45/90/0/90/-45/90/45/0/45//0/45/90/-45/90/-45/45/-45/0/90/0/45/0/45/-45/45/90/0/90/-45/90/-45/0/45]. The symbol '//' refers to the mid-plane location where a 15-μm Teflon film was placed to initiate the pre-crack. This provided the fibre orientation of 45°//0°adjacent to the delaminated surface. This stacking sequence led to uncoupled quasi-isotropic quasi-homogeneous (QIQH) property which was not only for the whole laminate, but also for each arm of the specimens [10]. For any 24-ply arm used here, the stiffness matrices A_{ij}, D_{ij} and B_{ij} were always the same and independent of the loading direction. Even if the laminate was not symmetric, the coupling stiffness matrix **[B]** and the entries A_{16}, A_{26}, D_{16} and D_{26} in the extensional **[A]** and bending **[D]**

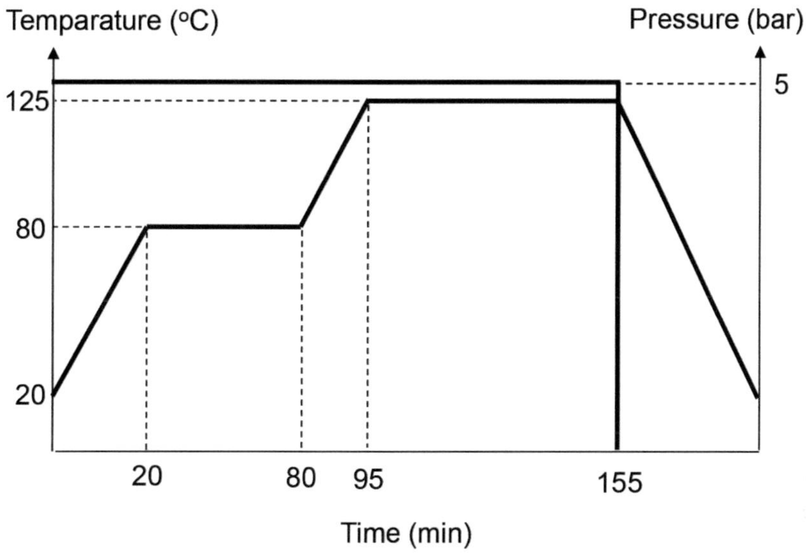

Figure 1. Curing cycle of the composite laminate fabrication.

stiffness matrices could be eliminated. Moreover, for other entries, A_{ij} was proportional to D_{ij}. This enabled random selection of fibre orientation in the layers adjacent and sub-adjacent to the interface crack. Besides, the non-dimensional ratio $D_{12}^2/(D_{11}D_{22})$ was always equal to 0.1036.

The composite was cured by using a hot-press machine according to the temperature cycle shown in **Figure 1**. Upon the completion of fabrication, the laminate was left to cool down under ambient condition.

3. Delamination tests

The composite plate was cut into specimens of 20 mm width using diamond-coated abrasive cutting blade with coolant. Three different types of fracture tests were conducted, which were double cantilever beam (DCB), 3-point end-notched flexure (ENF) and mixed-mode flexure (MMF) to characterise mode I, mode II and mixed-mode I+II delamination behaviour, respectively. The test configurations are illustrated in **Figure 2**. The composite laminate has the total thickness, $2h = 9.6$ mm. For both ENF and MMF tests, the half-span length, L, was always set to be 60 mm. The initial crack lengths, a_o, were 40 mm for DCB specimens and 25 mm for ENF and MMF specimens. The compliances of the specimens were also measured in the range of 45–70 mm for DCB specimens and 15–55 mm for ENF and MMF specimens. This was to obtain the compliance plots for the calculation of the fracture toughness. All tests were conducted using imposed crosshead speed of 1 mm/min. At least three replicates were tested for each series of specimens. All tests were conducted under ambient condition.

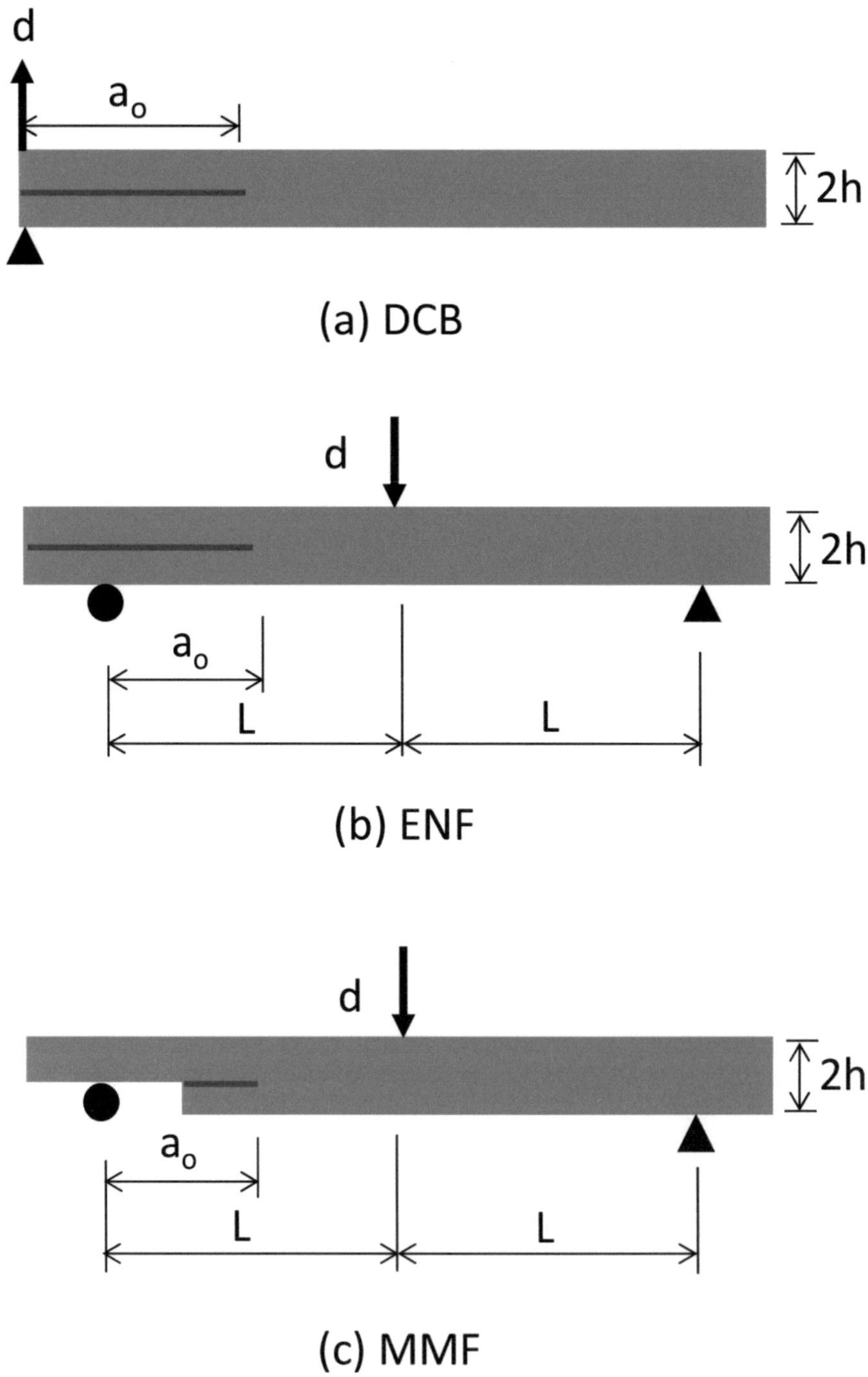

Figure 2. Test configurations for (a) DCB, (b) ENF and (c) MMF tests.

4. Data reduction scheme

The analysis method for the fracture energy was based on the Irwin-Kies equation [11], which is written as

$$G_C = \left(\frac{P_C^2}{2B}\right)\left(\frac{dC}{da}\right) \quad (1)$$

where G_C is the fracture energy, P_C is the critical load attained before propagation of crack initiation, B is the width of the specimen, C is the compliance and a is the crack length. For the experimental calibration method (ECM), the empirical compliance calibrated model is expressed by

$$C = C_2 a_0^3 + C_1 \quad (2)$$

where C_2 and C_1 are obtained from C versus $a_o{}^3$ plot. Substituting the derivative of Eq. (2) into Eq. (1) yields the following expression:

$$G_C = \left(\frac{P_C^2}{2B}\right) 3C_2 a_0^2 \quad (3)$$

Eqs. (2) and (3) are also applied to the crack propagation to determine the effective crack length, a_p and the resistance to crack growth in terms of the strain energy release rate, G_p, which are described as follows:

$$a_p = \left(\frac{C_p - C_1}{C_2}\right)^{\frac{1}{3}} \quad (4)$$

$$G_p = \left(\frac{P_p^2}{2B}\right) 3C_2 a_p^2 \quad (5)$$

where C_p and P_p signify the measured specimen's compliance and the load corresponding to an effective crack length, a_p. R-curves can then be obtained in order to investigate the crack propagation behaviour.

5. Force-displacement curves

Figure 3 displays the force-displacement curves of the specimens tested using DCB, ENF and MMF tests. All tests showed initial linear region, followed by load drop after peak load (critical load) was attained. The crack propagation in the DCB specimens was comparatively less stable compared to the ENF and MMF specimens. In addition, good repeatability was observed in all tests.

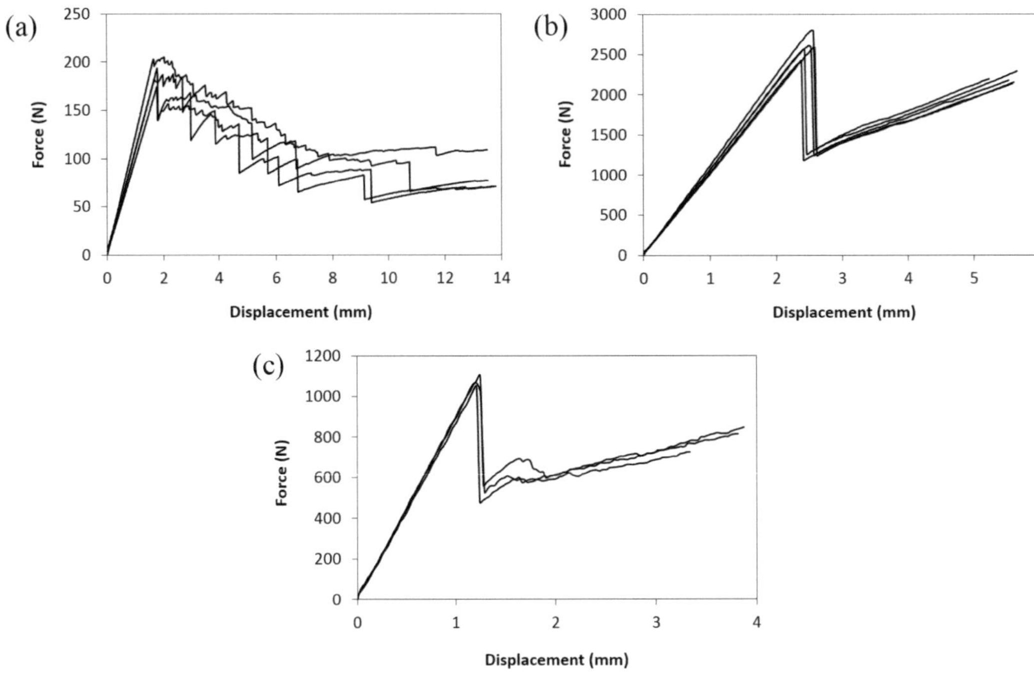

Figure 3. Force-displacement curves of (a) DCB, (b) ENF and (c) MMF specimens.

6. Mode I, mode II and mixed-mode I+II fracture toughness

The compliance plots of DCB, ENF and MMF tests are shown in **Figure 4**. By fitting the compliance data with a linear equation, C_2 and C_1 for each test were obtained and are shown in the figure. Subsequently, the fracture toughness could be calculated.

Figure 5 shows the fracture toughness at all three different mode ratios, where the values in the bracket refer to coefficient of variation (CV). The largest CV was less than 10%, which signified good repeatability of the tests. The variation of fracture toughness with respect to mixed-mode ratio was characterised using the Benzeggagh-Kenane (BK) mixed-mode criterion [12], which is expressed by Eq. (6):

$$G_{TC} = G_{IC} + (G_{IIC} - G_{IC})\left(\frac{G_{II}}{G_I + G_{II}}\right)^{\eta} \tag{6}$$

In the above equation, G_{TC} refers to the total fracture toughness for mixed-mode loading and η is the material constant and has to be determined empirically. From **Figure 5**, it could be noted that the fracture toughness increased with the mode ratio. The mode ratio, G_{II}/G_T of DCB, ENF and MMF are 0, 1 and 0.43, respectively. In addition, the best-fit η was 1.21, which was similar to the value reported for woven E-glass/bismaleimide [4], carbon/epoxy [8] and SL-EC glass-cloth/epoxy [13] composites.

http://dx.doi.org/10.5772/intechopen.69440

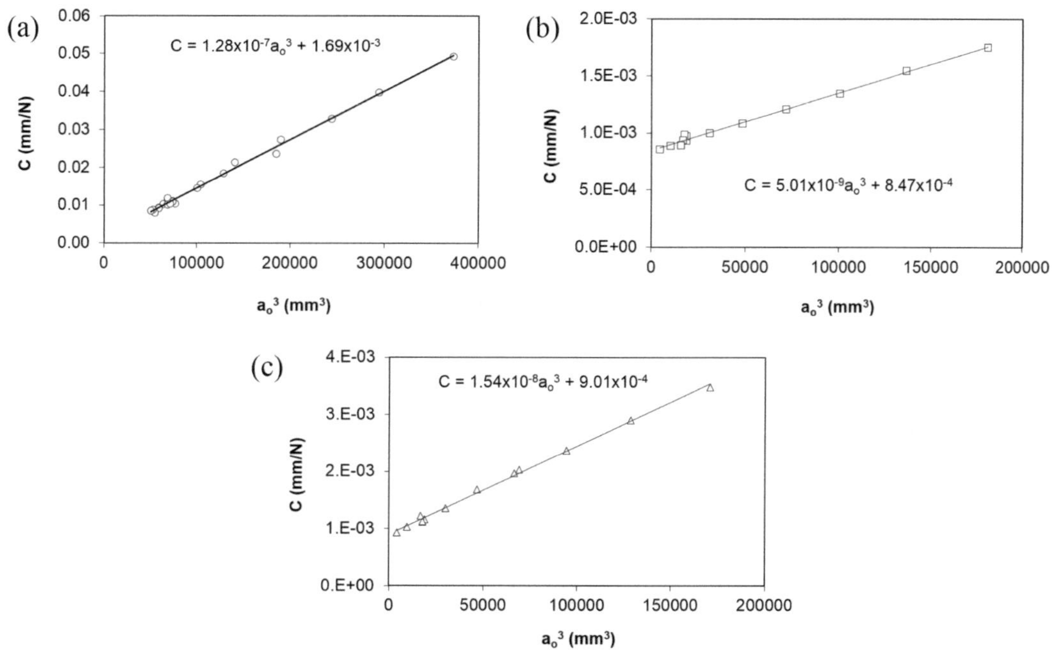

Figure 4. Compliance plots of (a) DCB, (b) ENF and (c) MMF specimens.

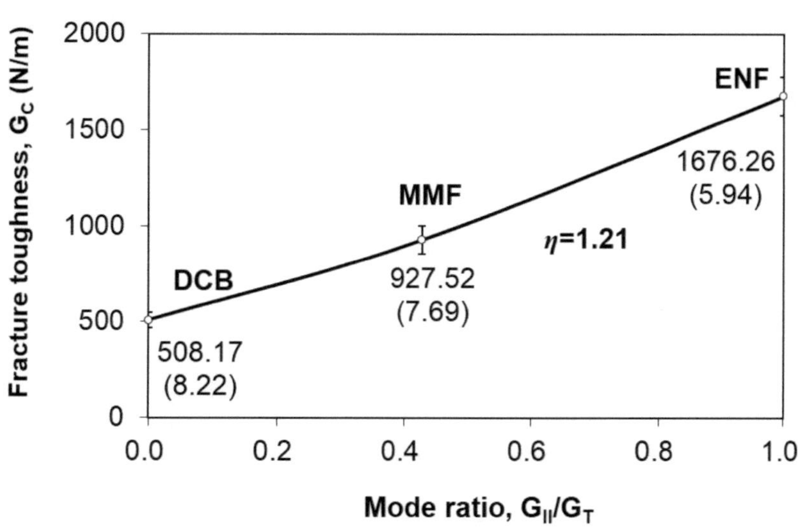

Figure 5. Variation of the fracture toughness with the mode ratio.

7. R-curve behaviour in mode I delamination

Figure 6 shows the R-curve of the DCB specimens. As shown in **Figure 6**, $da = a_p - a_o$ and $dG = G_p - G_C$, where a_p and G_p were calculated using Eqs. (4) and (5), respectively. The gaps between

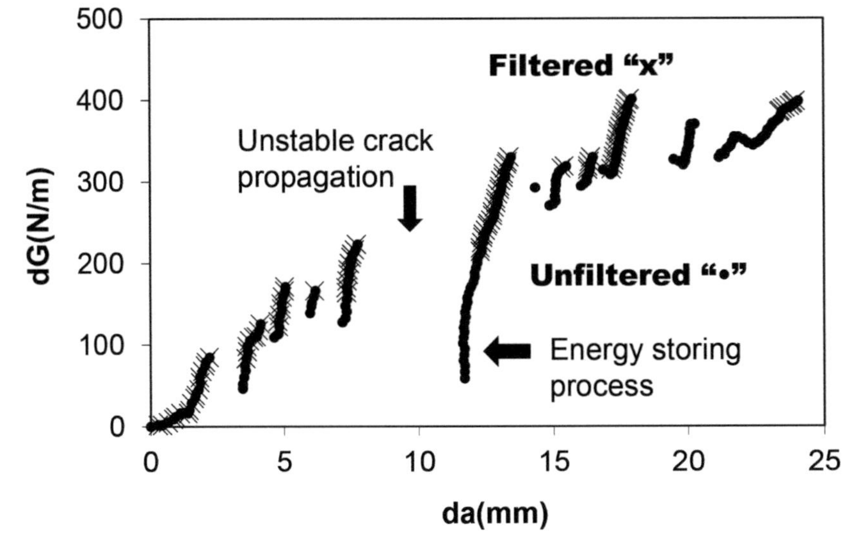

Figure 6. R-curve of the DCB specimens.

the data implied the crack jump due to instable crack propagation, where there was a sudden drop in the fracture energy. The energy magnitude of lower than the previous attained maxima did not contribute to additional crack propagation. Rather, it was an energy storing process and hence was filtered out. In this study, a drop in fracture energy of more than 5% from the previous maximum value was chosen as a basis for establishing representative R-curves. The filtered R-curve is marked as "x" in **Figure 6**. It was noted that the fracture energy increment was significant, where the maximum increment in the fracture energy (≈400 N/m) was approximately 80% of the G_{IC} value.

8. Fractographic analyses

Figures 7–9 show the comparison of the scanning electron micrographs of the delaminated surfaces of DCB, ENF and MMF specimens at the crack-tip region. It is obvious that there was significant resin-rich region at the crack tip of DCB and MMF specimens. In addition to the resin-rich region, fibre debonding was also found to be relatively significant in MMF specimen. As for the ENF specimen, matrix cracking and fibre debonding were found to be the two major damage mechanisms. This suggested that the fracture toughnesses that were calculated at different modes were actually contributed by the energy dissipation by different damage mechanisms.

To further investigate the delamination behaviour in the delamination region, additional scanning electron micrographs were captured and are shown in **Figures 10–12**. It was observed that for the DCB specimen, the delamination was dominated by the 45° layer, where fibre debonding was the major failure mechanism. Even on the 0° surface, traces of 45°layer

http://dx.doi.org/10.5772/intechopen.69440

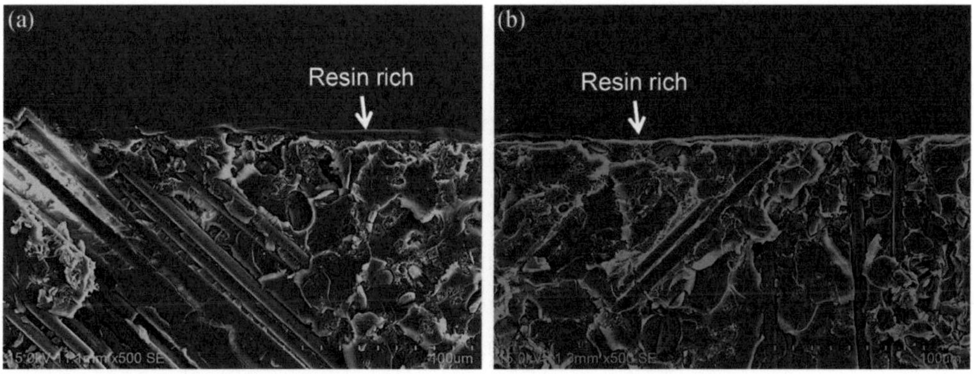

Figure 7. Crack-tip delamination of the DCB specimen: (a) 45°layer and (b) 0°layer.

Figure 8. Crack-tip delamination of the ENF specimen: (a) 45°layer and (b) 0°layer.

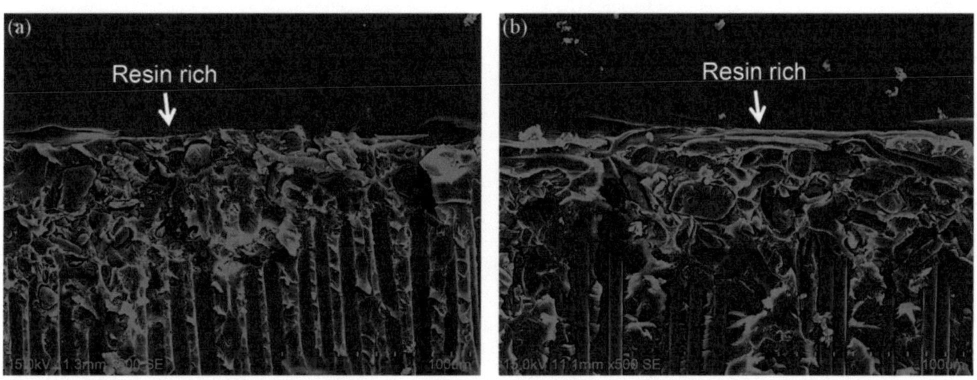

Figure 9. Crack-tip delamination of the MMF specimen: (a) 45°layer and (b) 0°layer.

were observed. Based on the R-curve behaviour observed, it was thus believed that fibre debonding has led to fibre bridging that enhanced the delamination growth and dissipated additional fracture energy [14].

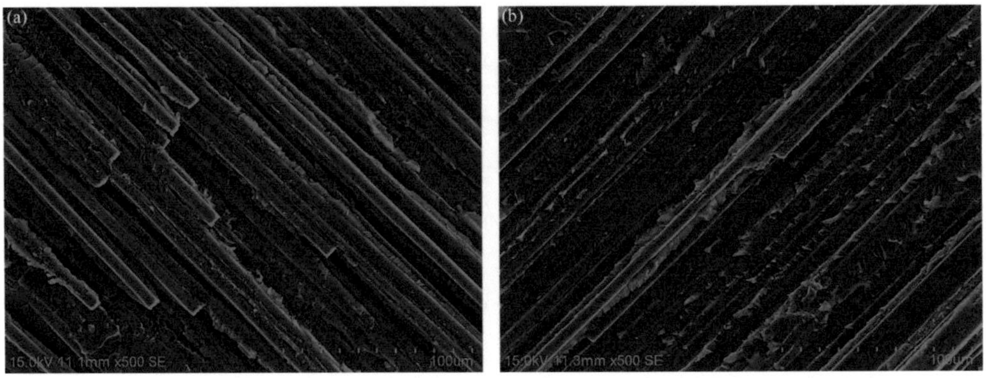

Figure 10. Delamination surfaces of the DCB specimen: (a) 45° layer and (b) 0° layer.

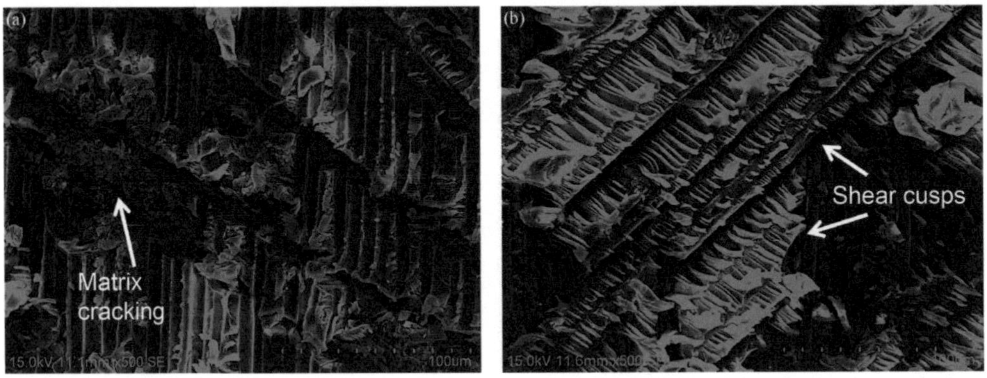

Figure 11. Delamination surfaces of the ENF specimen: (a) 45° layer and (b) 0° layer.

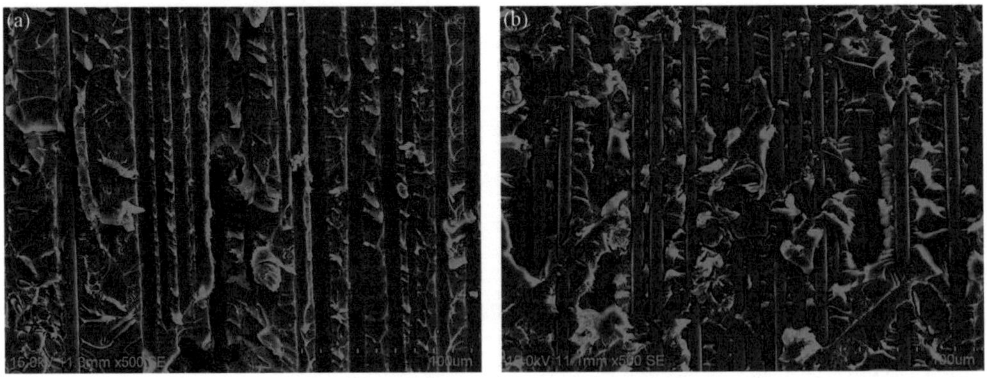

Figure 12. Delamination surfaces of the MMF specimen: (a) 45° layer and (b) 0° layer.

As for the ENF specimen, a significant difference in the delamination behaviour was observed. Fibre debonding was observed on the fibres in the 0°direction. However, significant shear cusps were observed on the 0°layer in the 45° direction, which was attributed to the shear mode loading. It has also been reported that the high shear mode would draw large shear cusps [6, 7]. A certain amount of matrix cracking was observed on the 45° layer, which was believed to be due to the interaction between 0°and 45° layers during shearing. Matrix hackles were also observed in the ENF specimen of a woven E-glass/bismaleimide composite [4].

As for the mixed-mode delamination, the MMF specimen exhibited fibre debonding dominated by the 0° layer, which was similar to the ENF specimen. Nevertheless, the shear cusps in the MMF specimen were less significant and more random compared to the ENF specimen, which was similar to the observation by Naghipour et al. [6]. In addition, the roughness of the MMF specimen was less than the ENF specimen [7]. On the other hand, matrix cracking seemed to be another major failure mechanism. All these observations suggested that the modes of loading influenced the failure mechanisms of the specimens, which in turned reflected in different values of fracture toughness.

9. Conclusion

In order to evaluate the delamination behaviour of a multidirectional laminated composite with 45° and 0° layers adjacent to the pre-crack, a quasi-isotropic quasi-homogeneous (QIQH) composite laminate was fabricated. Double cantilever beam (DCB), end-notched flexure (ENF) and mixed-mode flexure (MMF) tests were carried out to characterise mode I, mode II and mixed-mode I+II delamination, respectively. All tests were conducted at a constant crosshead speed of 1 mm/min under ambient condition. The average fracture toughnesses were mode I G_{IC} = 508.17 N/m, mode II G_{IIC} = 1676.26 N/m and mixed-mode I+II $G_{(I+II)C}$ = 927.52 N/m. The best-fit material parameter, η = 1.21, according to the Benzeggagh-Kenane (BK) criterion. Through scanning electron micrographs, the crack-tip region was found to be resin-rich. In addition, fibre debonding of the 45° layer was dominant in the DCB specimen, which was believed to have led to fibre bridging behaviour. Furthermore, in the ENF specimen, significant shear cusps were observed. Finally, matrix cracking and fibre debonding of the 0° layer were observed in the MMF specimen.

Acknowledgements

This work is supported by Ministry of Higher Education Malaysia and Universiti Teknologi Malaysia through Fundamental Research Grant Scheme (FRGS) No. 4F591. The authors also acknowledge Institut Supérieur de l'Automobile et des Transports (ISAT), France for providing the facilities for fabrication and testing.

Author details

Mahzan Johar, King Jye Wong* and Mohd Nasir Tamin

*Address all correspondence to: kjwong@mail.fkm.utm.my

Faculty of Mechanical Engineering, Universiti Teknologi Malaysia, Johor Bahru, Malaysia

References

[1] Soutis C. Fibre reinforced composites in aircraft construction. Progress in Aerospace Sciences. 2005;**41**(2):143-151. DOI: 10.1016/j.paerosci.2005.02.004

[2] Soutis C. Carbon fiber reinforced plastics in aircraft construction. Materials Science and Engineering A. 2005;**412**(1-2):171-176. DOI: 10.1016/j.msea.2005.08.064

[3] Turon A, Camanho PP, Costa J, Davila CG. A damage model for the simulation of delamination in advanced composites under variable-mode loading. Mechanics of Materials. 2006;**38**(11):1072-1089. DOI: 10.1016/j.mechmat.2005.10.003

[4] Zhao Y, Liu W, Seah LK, Chai GB. Delamination growth behavior of a woven E-glass/bismaleimide composite in seawater environment. Composites Part B: Engineering. 2016; **106**:332-343. DOI: 10.1016/j.compositesb.2016.09.045

[5] Pagano NJ, Schoeppner GA. Delamination of polymer matrix composites: Problems and assessment. In: Kelly A, Zweben C, editors. Comprehensive Composite Materials. Oxford: Pergamon; 2000. pp. 433-528. DOI: 10.1016/B0-08-042993-9/00073-5

[6] Naghipour P, Schneider J, Barsch M, Hausmann J, Voggenreiter H. Fracture simulation of CFRP laminates in mixed mode bending. Engineering Fracture Mechanics. 2009;**76**(18): 2821-2833. DOI: 10.1016/j.engfracmech.2009.05.009

[7] Naghipour P, Bartsch M, Chernova L, Hausmann J, Voggenreiter H. Effect of fiber angle orientation and stacking sequence on mixed mode fracture toughness of carbon fiber reinforced plastics: Numerical and experimental investigations. Materials Science and Engineering A. 2010;**527**(3):509-517. DOI: 10.1016/j.msea.2009.07.069

[8] LeBlanc LR, LaPlante G. Experimental investigation and finite element modeling of mixed-mode delamination in a moisture-exposed carbon/epoxy composite. Composites Part A: Applied Science and Manufacturing. 2016;**81**:202-213. DOI: 10.1016/j.compositesa.2015.11.017

[9] Liu Y, Zhang C, Xiang Y. A critical plane-based fracture criterion for mixed-mode delamination in composite materials. Composites Part B: Engineering. 2015;**82**:212-220. DOI: 10.1016/j.compositesb.2015.08.017

[10] Vannucci P, Verchery G. A new method for generating fully isotropic laminates. Composite Structures. 2002;**58**(1):75-82. DOI: 10.1016/S0263-8223(02)00038-7

[11] Irwin GR, Kies JA. Critical energy rate analysis of fracture strength. Welding Journal Research Supplement. 1954;**33**:193-198

[12] Benzeggagh ML, Kenane M. Measurement of mixed-mode delamination fracture toughness of unidirectional glass/epoxy composites with mixed-mode bending apparatus. Composites Science and Technology. 1996;**56**(4):439-449. DOI: 10.1016/0266-3538(96)00005-X

[13] Shindo Y, Shinohe D, Kumagai S, Horiguchi K. Analysis and testing of mixed-mode interlaminar fracture behavior of glass-cloth/epoxy laminates at cryogenic temperatures. Journal of Engineering Materials and Technology. 2005;**127**(4):468-475. DOI: 10.1115/1.2019944

[14] Barikani M, Saidpour H, Sezen M. Mode-I interlaminar fracture toughness in unidirectional carbon-fibre/epoxy composites. Iranian Polymer Journal. 2002;**11**(6):413-423

Failure Analysis of High Pressure High Temperature Super-Heater Outlet Header Tube in Heat Recovery Steam Generator

Ainul Akmar Mokhtar and
Muhammad Kamil Kamarul Bahrin

Additional information is available at the end of the chapter

http://dx.doi.org/10.5772/intechopen.72116

Abstract

Heat Recovery Seam Generator (HRSG) tube failure is one of the most frequent causes of power plant forced outage. In one of the local power plants, one of the boilers has experienced several defects and failures after running approximately 85,000 hours. 17 tube failures were found at the High Pressure High Temperature Superheater (HPHTSH) outlet header. The aim of this study is to find the root cause of the tube failures and to suggest the remedial action to prevent repetitive failure event. Several analysis methods were conducted to ascertain the potential cause(s) of failure. The results showed that the tubes failed due to long-term creep and thermal fatigue based on the cracking behaviour. Furthermore, the power plant has been operating as a peaking plant which concluded that the tubes have undergone the thermal stress due to frequent temperature change in the tubes. Flow correcting device (FCD) was also found damaged, causing flow imbalance in the tubes. Flow imbalance accelerated the creep degradation on the tubes. It was recommended that the FCD has to be repaired and improved to balance the flow. Furthermore, the extensive life assessment was recommended to be done on all the tubes to avoid future tube failures.

Keywords: boiler tube failure, HRSG, failure analysis

1. Background of study

After nearly 85,000 running hours, one of the boilers, also known as heat recovery steam generator (HRSG), in one of the local power plants has been experiencing 17 boiler tube failures. The power plant was designed as base load plant; however, after 7 years of operation, the plant started changing its operating regime from base load to cyclic load with more

INTECH
open science | open minds

© 2017 The Author(s). Licensee InTech. This chapter is distributed under the terms of the Creative Commons Attribution License (http://creativecommons.org/licenses/by/3.0), which permits unrestricted use, distribution, and reproduction in any medium, provided the original work is properly cited. (cc) BY

start-up/shutdown. The frequency of tube failures has increased when the plant was having more cyclic operation as compared to 5 years after commissioning. The connecting tubes of the outlet header of HRSG High Pressure High Temperature Superheater (HPHTSH) have suffered leakage, and all the failed tubes were located next to each other in a row (i.e. Row 3) as shown in **Figure 1**.

During the incident, a significant drop of the water level and water pressure was detected at the high pressure (HP) boiler drum due to the extensive leak of these HPHTSH connecting tubes. Then, the plant decided to declare force outage in order for inspection team to carry out proper, thorough inspections to identify the cause of water loss. The suspected tubes of High Temperature Reheat (HTRH), HPHTSH and High Pressure Low Temperature Superheater (HPLTSH) inlet and outlet header were inspected.

The outcomes from the inspection showed that no defects were found on HTRH and HPLTSH inlet and outlet header; however, circumferential cracks were found at 17 tubes on HPHTSH outlet header. The flow correcting devices (FCD) also were found detached and misaligned from its original position. Maintenance team has appointed internal inspection team to investigate the possible failure mechanism and the root cause of HPHTSH outlet header tube failures. Also, they requested external laboratory to conduct a remaining creep life prediction [1].

The main objectives are:

1. To determine metallurgical characteristics and mechanical properties of the failed tube specimens.
2. To identify the root cause of failures and its mechanisms.
3. To provide recommendations for rectification actions in future prevention of the tube failures.

The investigation covered the failure of Tube 8, Tube 9, Tube 10 and Tube 18 of the HPHTSH section. **Figure 2** shows the samples of tube failures. Several analysis methods were carried out to ascertain the potential cause(s) of failure such as visual examination, dimensional

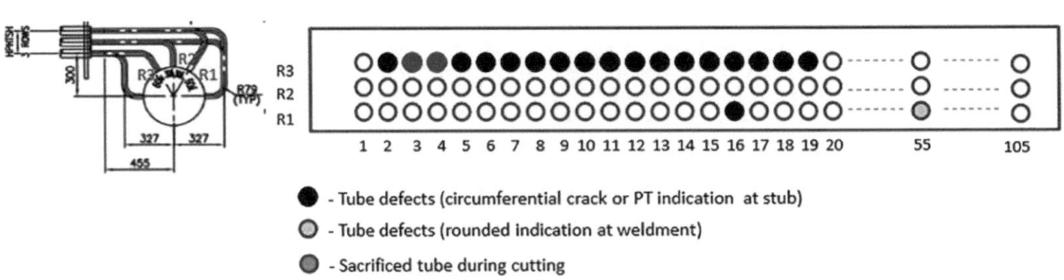

Figure 1. Tube mapping of recent failure of HPHTSH outlet header tubes.

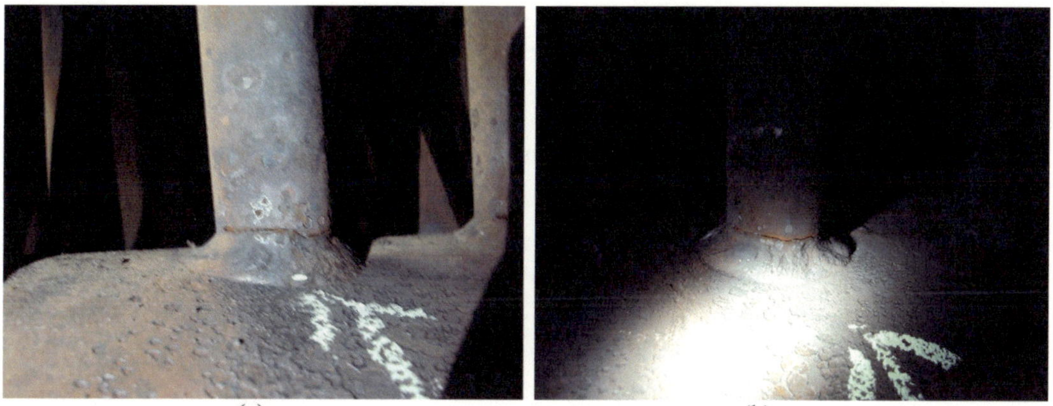

(a) (b)

Figure 2. Appearance of Row 3 tube failures, (a) Tube 8 and (b) Tube 9.

Type of analysis	Tube 8	Tube 9	Tube 10	Tube 18
Visual inspection	√	√	√	√
Dimensional measurement	√			√
Metallographic examination	√			√
Hardness testing	√			√
XRF analysis	√	√	√	√
Oxide thickness assessment	√			√

Table 1. Analysis done on each tube.

measurement, metallurgical analysis, hardness testing, chemical analysis or X-ray fluorescence (XRF) analysis, oxide layer measurement and creep assessment. Each method of analysis may not cover all tubes due to limited resources and time. **Table 1** shows the analysis method done on those selected tubes.

2. Procedure for HRSG tube failure analysis

Detail visual examination is the first step to determine earlier condition of failed tube. After received the sample, the condition on fireside and waterside surface shall be thoroughly inspected for any abnormal indication and the findings shall be recorded with photographic documentation. Any change of colour, deposit texture, fracture surface, location of defect and morphology should be focussed and recorded.

The next step in the tube failure analysis is to conduct a dimensional analysis. The design data and drawing should be available before proceeding this step. Quantitative assessment on failed tube with the Vernier calliper or ID/OD micrometre useful in assessing wall thinning, bulging and any corrosion damage. Ultrasonic thickness machine can be used to measure the oxide thickness of failed area and give the idea how long the failure has already taken place. The extent of oxide formation and/or ductile expansion can provide an idea to determine the primary failure mechanism. For example, thick oxide may indicate that the rupture may be due to overheating.

The appearance and/or orientation of crack can be helpful in determining a failure mechanism. While A soot blower erosion typically causes 'fish mouth' and thick-edged appearance, and thermal commonly results in transverse crack at heat-affected zone (HAZ). Normally, the crack shall be examined closely at HAZ and weldment area. Non-Destructive Test (NDT) such as dye-penetrant testing or magnetic particle testing may be necessary to determine the extent of the crack [2].

The evaluation of water chemistry and boiler operation is necessary to determine the failure mechanism. As an example, poor water chemistry will result in flow accelerated corrosion (FAC) on low pressure section or thicker oxide layer in a superheated section. Excessive oxide layer will result in high metal temperature. This will eventually cause high-temperature oxidation and may lead to exfoliation. Quantitative analysis of the internal tube surface commonly involves the determination of the oxide scale value and deposit thickness. Interpretation of these values can define the role of internal deposits in a failure mechanism. Oxide scale values are also used to determine whether chemical cleaning of boiler tubing is required. Wet chemical analysis is often used to determine the elemental composition. A number of spectrochemical analysis techniques can also be used for the quantitative measurement to determine the material percentage in the failed tube. The result will be compared to ASME, 2015, Section II Part D. In boiler construction, there are several cases where the installation or repair work was done using wrong grade of materials. In any cases, wet chemical analysis comparison of the steel can be used to determine the cause of premature failure.

Hardness measurement can be used to estimate the tensile strength/material hardness. The comparison of failed tube harness to accepted hardness range, and the deterioration of the mechanical properties can be determined due to material creep. The micro-Vickers hardness tester is usually used to obtain the hardness profile along the defective surface including HAZ, weldment, and base metal for any microstructure change on the material.

Metallography is key actions in determining the boiler tube failure mechanism. It is also good in assess the following:

a. whether cracks initiated on a waterside or fireside surface

b. the orientation of crack

c. whether a tube failure resulted from hydrogen embrittlement

d. whether a tube failed from short-term or long-term overheating damage

e. whether cracks were caused by creep damage, thermal fatigue, or stress-corrosion cracking.

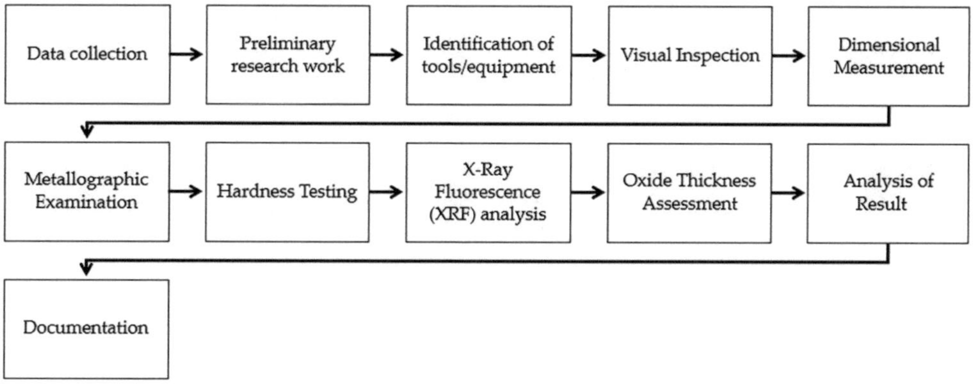

Figure 3. Flow chart of failure analysis activities.

Proper sample orientation and preparation are critical aspects of microstructural analysis. The orientation of the sectioning is determined by the specific failure characteristics of the case. After careful selection, metal specimens are cut with an abrasive cut-off wheel and mounted in a mould with resin or plastic. After mounting, the samples are subjected to a series of grinding and polishing steps. The goal is to obtain a flat, scratch-free surface of metal in the zone of interest. After processing, a suitable etchant is applied to the polished metal surface to reveal microstructural constituents (grain boundaries, distribution and morphology of iron carbides, etc.). **Figure 3** summarizes the failure analysis activities.

3. Results and discussion

3.1. Visual examination

Four tube sections were identified in Row 3 as Tube 8, Tube 9, Tube 10 and Tube 18, and each marked with an arrow indicating the flow direction towards the outlet header. Measurement has showed that the five tube sections have a nominal outer diameter (OD) of 32 mm. All examined tubes comprised bend sections approximately 280 mm long, measured around the extrados of the bend.

The connecting tubes are welded directly on the header without the presence of stubs. Cracks have developed at the heat-affected zone (HAZ) of the welds on the tube side. All cracks were circumferential in nature with no branching characteristics. The cracks propagated at the outer radius plane of the connecting tubes with an average length of approximately 25 mm as shown in **Figure 4**.

The tube sections have not undergone visible bulging and deformation over the entire length. The external surfaces of all sections appeared to be normal, with no signs of significant corrosion/erosion damage. The magnetite layer on the inner, steam-touched surfaces appeared thick, homogenous, adhesive and continuous, with no signs of serious/apparent exfoliation. The circumferential cracks have penetrated the tube wall.

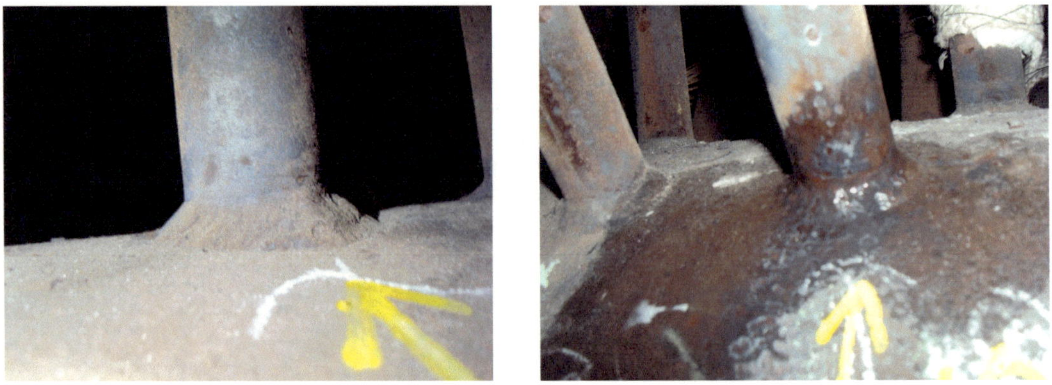

(a) (b)

Figure 4. Appearance of tube failure at Row 3: (a) Tube 8 and (b) Tube 18.

3.2. Dimensional measurement

3.2.1. Thickness measurement

Tube 8 and Tube 18 out of five tubes were chosen for dimensional measurement since the Tube 8 may be a representative of Tube 9 and Tube 10. The tube was cut into six pieces in order to do OD and inner diameter (ID) measurement. The thickness of Tube 8 and Tube 18 was measured by ultrasonic thickness at three locations for each tube, designated as A, B and C. Four spots with respect to clock position 12, 3, 6 and 9 were measured at each location as shown in **Figure 5**. The results are recorded in **Tables 2** and **3**.

The minimum measured thickness readings were compared with the minimum required thickness from the RBI report and Doosan (HRSG OEM) report. Several locations of Tube 8 and Tube 18 showed that the wall thickness was lower than the minimum required thickness, which was considered unsafe and unacceptable. It implied that the risk of stress rupture

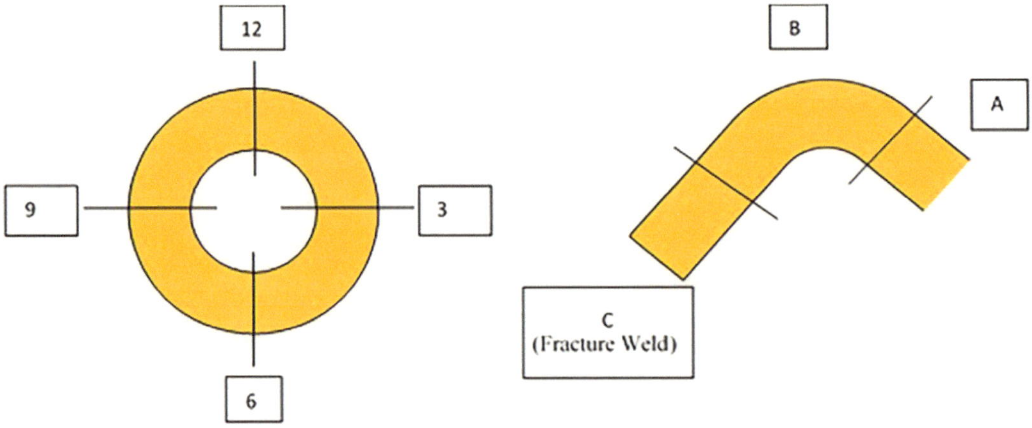

Figure 5. Dimensional references of thickness measurement on a tube.

Tube 8				
Location	**12**	**3**	**6**	**9**
A	2.95	2.79	2.86	2.87
	2.96	2.7	2.84	2.83
	2.95	2.76	2.87	2.81
Min value	2.95	2.7	2.84	2.81
B	2.14	2.96	3.28	2.83
	2.16	2.93	3.29	2.89
	2.15	2.95	3.24	2.85
Min value	2.14	2.93	3.24	2.83
C	2.88	2.73	2.83	2.92
	2.81	2.74	2.82	2.9
	2.82	2.73	2.8	2.91
Min value	2.81	2.73	2.8	2.9

Table 2. Thickness reading of Tube 8 (in mm).

was high. The significant wall thinning could be due to the excessive steam oxidation, which caused a high consumption rate of metal.

3.2.2. Outer/inner diameter (OD/ID) measurement

Tube 8 and Tube 18 were cut at an interval length of 50 mm. The sections are designated 1–6 as shown in **Figure 6**. The OD, ID and thickness of each cutting location were measured by Vernier calliper at the area shown in **Figure 7**. The results are recorded in **Tables 4** and **5**.

Tube 18				
Location	**12**	**3**	**6**	**9**
A	2.77	2.63	2.89	2.88
	2.74	2.65	2.91	2.89
	2.75	2.62	2.9	2.83
Min value	2.74	2.62	2.89	2.83
B	2.21	2.86	3.41	2.86
	2.2	2.82	3.43	2.89
	2.19	2.91	3.42	2.81
Min value	2.19	2.82	3.41	2.81
C	2.72	2.72	2.7	2.71
	2.79	2.73	2.78	2.69
	2.74	2.71	2.76	2.72
Min value	2.72	2.71	2.7	2.69

Table 3. Thickness reading of Tube 18 (in mm).

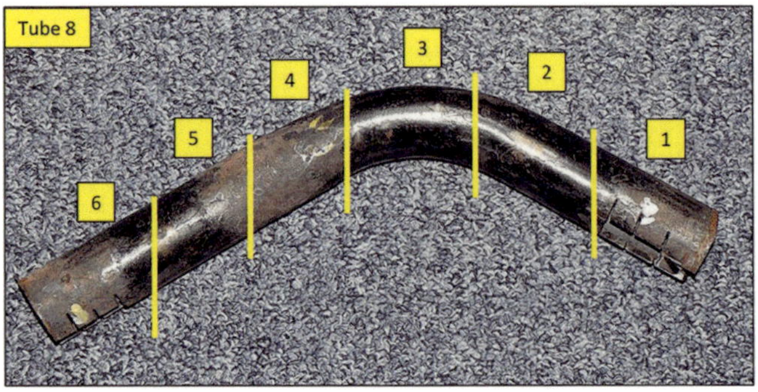

Figure 6. Tube 8 was cut into six sections.

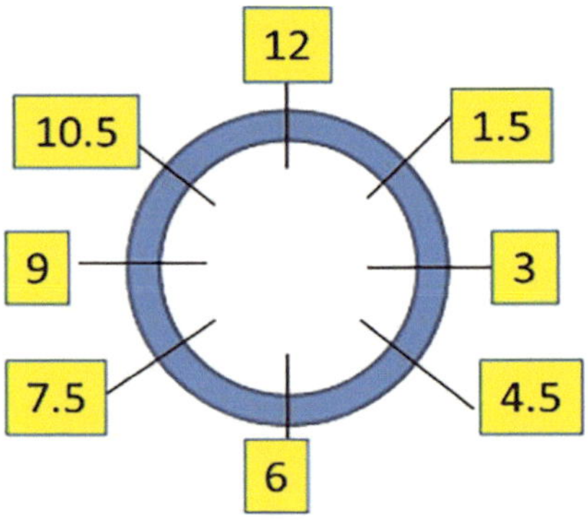

Figure 7. Dimensional reference of OD and ID tube.

The OD and ID for both tubes did not show any significant signs of tube bulging, deformation and wall thinning over the length, as compared to the design values, where OD was 31.8 mm and ID was 26.2 mm.

3.3. Metallographic examination

The metallographic examination was performed on Tube 8 and Tube 18 through the replication method and the standard metallographic preparation method. The replication method was applied on tube surface in a non-destructive manner, while in the standard metallographic method, the cracked part was cut and mounted in cross sections. The damage characteristics of the fractured area in mounted specimens of Tube 8 and Tube 18 are described as follows:

a. The cracks were fairly straight and had a transgranular appearance, thick oxide filled, without branching characteristics. The cracks propagated in the HAZ.

b. Isolated, aligned cavities and microcracks were presented in the vicinity of the cracks. The cavities were predominantly aligned in parallel to the cracks, and some were perpendicular to the cracks.

c. Under high magnifications, the grade 91 microstructure of base metal still appeared normal (martensitic) with no signs of thermal degradation such as phase transformation, grain growth and precipitate coarsening. A small amount of isolated cavities were found in the remote base metal of Tube 8.

d. The fracture surfaces had oxidized heavily. The oxide layers at the steam touch surfaces (inner surface) were considered thick, in excess of 300 μm.

Figure 8 shows the microstructure of Tube 8 at the fracture area at different magnification. Isolated cavities, aligned cavities, microcrack and thick oxidation layer were observed at that particular location.

3.4. Hardness testing

Micro-Vickers hardness was used to measure the mounted specimen hardness of Tube 8 and Tube 18. The applied load for this test was 10 kg load in 5 s. The tested locations include:

a. Heat-affected zone (HAZ)

b. Base metal (near crack's location)

c. Base metal (remote area)

Section	Outer diameter (mm)				Inner diameter (mm)			
	12–6	1.5–7.5	3–9	4.5–10.5	12–6	1.5–7.5	3–9	4.5–10.5
1	31.75	31.89	31.82	31.87	25.25	24.47	25.31	25.31
2	31.90	31.86	31.86	31.87	24.80	24.98	25.22	25.34
3	31.93	30.75	30.65	31.85	25.16	24.11	23.89	24.63
4	31.85	32.03	31.94	31.98	24.63	25.18	25.11	24.84
5	31.88	31.76	31.77	31.87	25.20	25.16	25.55	25.26
6	31.66	31.71	31.60	31.43	25.14	25.11	24.64	24.67

Table 4. OD and ID each section of Tube 8.

Section	Outer diameter (mm)				Inner diameter (mm)			
	12–6	1.5–7.5	3–9	4.5–10.5	12–6	1.5–7.5	3–9	4.5–10.5
1	31.69	31.25	31.57	31.29	25.29	25.12	25.35	25.28
2	31.52	31.74	31.87	30.82	24.84	24.67	25.40	24.17
3	31.31	31.35	30.74	32.22	24.56	23.95	23.84	25.27
4	30.97	30.59	29.51	31.63	24.80	23.78	23.14	25.04
5	32.02	32.31	31.91	32.22	25.18	25.14	25.03	25.33
6	31.86	31.63	31.60	31.60	25.04	24.89	25.35	25.12

Table 5. OD and ID each section of Tube 18.

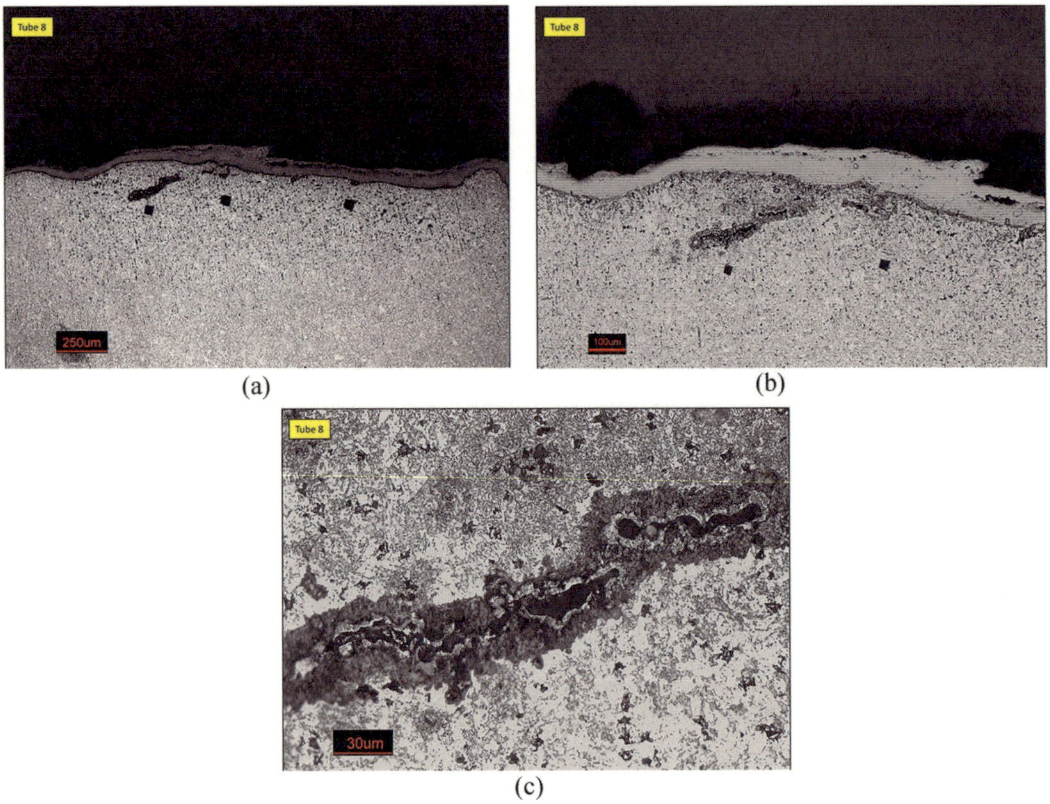

Figure 8. The microstructure of Tube 8 shows a crack at different magnification: (a) 250 μm, (b) 100 μm and (c) 30 μm.

The result of hardness testing is reported in **Table 6**.

Based on the ASTM 213 T91 code, the maximum hardness for tube material 91 is 265 HV [3]. Based on experience, the low alloy material 91 hardness should not be less than 195 HV. The base metal of the tested part was considered still within design specification. The HAZ has a low hardness property, which is considered not acceptable, especially 142 HV, which indicates progressive thermal degradation.

Location	Average hardness (HV)	
	Tube 8	Tube 18
Weld joint	206	267
Heat-affected zone (HAZ)	142	185
Base metal	212	220
Base metal	202	221

Table 6. Average hardness value of selected locations of Tube 8 and Tube 18.

3.5. X-ray fluorescence (XRF) analysis

Using a Niton XLt 898 XRF analyser, a positive material identification (PMI) analysis was carried out on the four tube sections (i.e. Tube 8, Tube 9, Tube 10 and Tube 18). Prior to the analysis, the external surface of each section was ground back to bright metal, and the analysis was carried out for three times. For the elements analysed, the average results obtained from both analyses on each section are summarized in **Table 7**, together with the range in composition specified for ASTM A213-09 Grade T91, shown for comparison.

Within the limits of the instrument used, the results of the XRF analysis showed that all four sections were manufactured using ferritic alloy steel, which conformed to Grade T91, that is the specified grade of material. One minor anomaly was the chromium contents of the R3 Tube 18, which contained 7.9% Cr, marginally below the minimum specified chromium content of 8.00%.

3.6. Operational assessment

3.6.1. Statistics of failed/repaired tube

The plant had experienced several HRSG tube failures for its HPHTSH section during its 12 years of operations. Statistics of the tube failures by locations are as per Failure 14, which shows that the majority of the failure has occurred at the right sidewall, followed by the left sidewall. Based on computational fluid dynamic (CFD) study done by third-party consultant (Andreas F., 2015), it was found that there are uneven flow and temperature distribution of the GAS turbine exhaust gas over the HPHTSH section. As shown by the flow study, during the base load operation of 350 MW, the flow is mainly concentrated at the two sidewalls while at the minimum load operation of 210 MW, the flow is channelled only to concentrate at the right-sidewall of the HRSG. Over its 12 years of operations, these operations had caused some tubes for the section to have higher tendency for creep damage as opposed to the others. The failures in the middle section were identified to be related to the acoustic baffle plates, which had been resolved in the past.

From the temperature trending, there are significantly more than six readings where the temperature drops below 100°C within 1-year period. The frequent start-stop, especially cold start, may accelerate thermal fatigue damage, for example, ligament cracking in stud locations (**Figure 9**).

Tube no	Element (wt%)					
	Chromium	Molybdenum	Nickel	Manganese	Vanadium	Niobium
Tube 8	8.40	1.06	0.11	0.52	0.20	0.08
Tube 9	8.10	0.97	0.07	0.52	0.21	0.08
Tube 10	8.40	1.02	0.06	0.49	0.23	0.08
Tube 18	7.90	0.97	0.10	0.55	0.23	0.08
ASTM A213 Grade T91	8.00/9.50	0.85/1.05	0.40 max	0.30/0.60	0.18/0.25	0.06/0.10

Table 7. The chemical composition of defective tube [3].

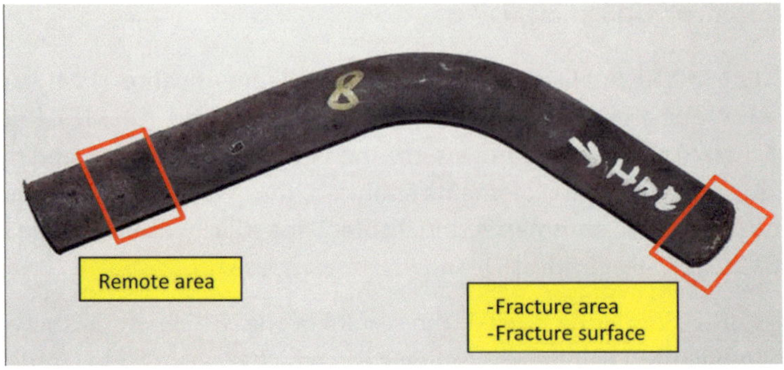

Figure 9. Dimensional reference of oxide thickness measurement.

3.7. Oxide thickness assessment

The location of interest for internal oxide assessment for Tube 8 and Tube 18 is shown in **Figure 9**.

3.8. Comparison

The steam-side oxide scale of micro specimens of Tube 8 and Tube 18 was measured under optical microscope. Both Tube 8 and Tube 18 were found to have undergone excessive steam oxidation. The oxide layer had cracked. The oxide thickness was measured and tabulated in **Figures 10** and **11**. The maximum thickness of the steam side oxide scale was 363 μm for Tube 8 and 390 μm for Tube 18.

Based from steam oxidation calculation [4], the thick oxide scale suggested that both Tube 8 and Tube 18 had been exposed to temperature in excess of 615°C as compared to material specification of SA 213T91 (design temperature, 602°C). The two tubes had experienced long-term overheating during operation.

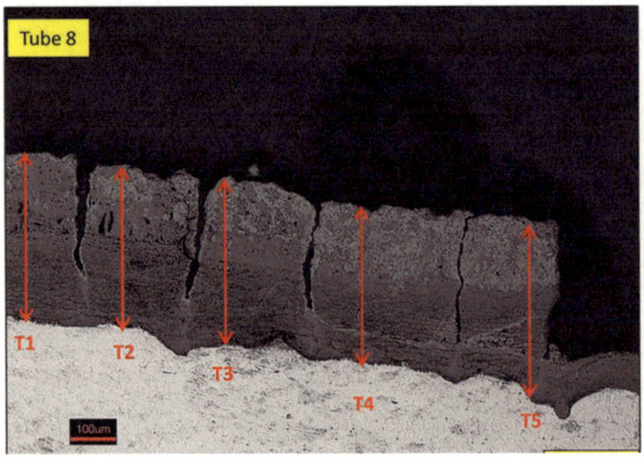

Points	Thickness (μm)
T1	369.44
T2	358.33
T3	361.11
T4	350.00
T5	377.78
Average	363.33

Figure 10. Internal oxide thickness at Tube 8 (fracture area).

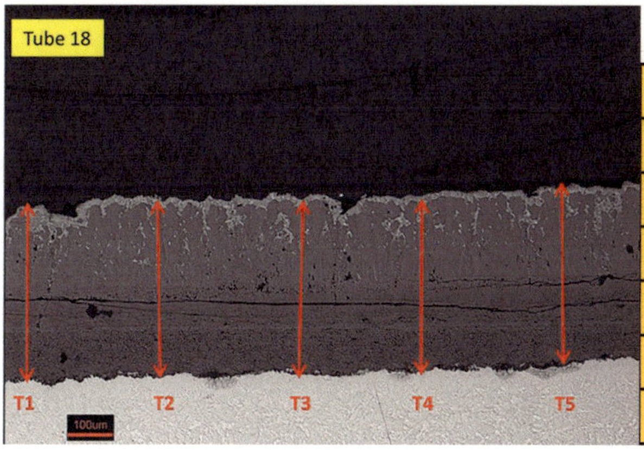

Points	Thickness (μm)
T1	394.44
T2	388.89
T3	388.89
T4	383.33
T5	397.22
Average	390.554

Figure 11. Internal oxide thickness at Tube 18 (remote area).

Tube	Average oxide thickness (μm)	Exposure temperature (°C)	Exposure hour
Tube 8 (internal fracture)	363	630.74	85,000
Tube 8 (internal remote)	270	610.4	85,000
Tube 8 (fracture surface)	118	630.74	8982
Tube 18 (internal fracture)	355	629.18	85,000
Tube 18 (internal remote)	322	622.4	85,000
Tube 18 (fracture surface)	107	629.18	7722

Table 8. The correlation between oxide thickness and exposure temperature/hour using oxide kinetic model.

From the steam oxide calculation, the oxide-filled fracture paths indicated that the cracks had existed for more than 7000 h. **Table 8** shows the oxide calculation based on oxide model.

Both the tubes (i.e. Tube 8 and Tube 18) have been exposed to temperatures of more than 600°C as compared to material specification of SA 213T91, where the design temperature is only 602°C. With this exposure, it had further promoted degradation of tubes material and formation of creep cavities. For fracture surface at Tube 8, with oxide thickness of 118 μm, the crack had been initiated for 8982 h. For fracture surface at Tube 18, with oxide thickness of 107 μm, the crack had been initiated for 7722 h.

3.9. Summary of analysis

The cracking showed the following characteristics:

a. Fairly straight

b. Mainly transgranular

c. No branching

d. Filled with thick oxides

e. Numerous aligned cavities in the vicinity of the cracks

The above observation suggested that the tubes had likely suffered a combination of two damage mechanisms, which were fatigue and short-term creep. The numerous and extensive creep cavities in the HAZ suggested that the tubes at the weld region had undergone advanced and progressive creep damage. For Tube 8, microcracks had formed and the creep damage was classified as 4 according to rating guideline provided in the NORDTEST NT TR 170 (1992) code. For Tube 18, the creep cavities were preferentially oriented and aligned, suitably to be classified as 3.3 according to the NORDTEST rating. The heavy thermal degradation was supported by the low hardness readings at the HAZ (minimum 142 HV was reported, as compared to acceptable range 200 HV–260 HV). It was obvious that the two tubes had undergone time-dependent, high-temperature creep process.

The plant had been operating as peaking plant, and it was subjected to daily start-stop. As the HAZ of the connecting tubes had experienced high-temperature creep degradation after 85,000 h had elapsed, the creep-fatigue cracks could have initiated at this weak region. The propagation was accelerated by the thermal cyclic stress, which was imposed by daily start-stop.

The creep mechanism could have been promoted by the following factors:

a. Defective FCD causing the instability of heat flux distribution, excessive metal temperature, and localized heating at the tubes (HPHTSH header metal temperatures were high, more than 602°C, the T91 limit).

b. Poor quality of HAZ (possibly caused by improper PWHT).

The thermal fatigue could have been promoted by thermal stress imposed by daily start-stop cycles.

4. Conclusions and recommendations

The cracks at the HAZ of the Row 3 connecting tubes at HPHTSH header could be attributed to creep-fatigue damage. The following contributory factors were identified:

1. Excessive metal temperature, as compared to design value.
2. Low hardness property at HAZ (creep strength was low).
3. Long service exposed hours (85,000 h–approaching 100,000 h).
4. Frequent start-stop cycles.
5. Localized heating due to instable heat flux distribution (contribute to creep)

Nevertheless, the remaining wall thickness had reduced to less than the minimum required thickness of 2.8 mm; the tubes were prone to stress-rupture failure, which might have occurred in short term. Some of the recommendations are given as follows:

1. As the connecting tubes had begun to show signs of advanced creep damage at HAZ, there is likelihood that the header, especially at the weld and HAZ had also undergone similar creep damage. It is recommended to conduct extensive replication and hardness testing at all the connecting tubes and header shell during outage opportunities. The area of interest is weld and HAZ.
2. The thickness profile of all the affected superheater tubes shall be measured in extensive manner during outage opportunities. Tube sections with remaining wall lower than minimum required thickness shall be cut and replaced.
3. As the plant is subjected to daily start-stop recently, it was recommended to conduct borescope inspection inside header to check for ligament cracking. It was also proposed to minimize the request from the single buyer in order to minimize the daily start/stop, thus, minimizing the cyclic operation and fatigue stress.
4. The plant should consider on conducting a comprehensive life assessment study on the boiler, as the service hours had approached 100,000 h and signs of creep damage had emerged.

Author details

Ainul Akmar Mokhtar*† and Muhammad Kamil Kamarul Bahrin†

*Address all correspondence to: ainulakmar_mokhtar@utp.edu.my

Universiti Teknologi PETRONAS, Perak, Malaysia

† These authors contributed equally.

References

[1] ASM International. Introduction to Steels and Cast Irons—Metallographer's Guide: Irons and Steels. Ohio: ASM International; 2002

[2] ASME. Section V – Non-Destructive Examination. ASME Boiler & Pressure Vessel Code. New York, NY: The American Society of Mechanical Engineers; 2015

[3] ASME. Section II Part D–Material Properties. ASME Boiler & Pressure Vessel Code. New York, NY: The American Society of Mechanical Engineers; 2015

[4] API Std 579-1/ASME FFS-1. Fitness-For-Service. Washington, DC: American Petroleum Institute; 2016

Chapter 5

Thick-Film Resistor Failure Analysis Based on Low-Frequency Noise Measurements

Ivanka Stanimirović

Additional information is available at the end of the chapter

http://dx.doi.org/10.5772/intechopen.69442

Abstract

The chapter aims to present research results in the field of thick-film resistor failure analysis based on standard resistance and low-frequency noise measurements. Noise spectroscopy–based analysis establishes correlation between noise parameters and parameters of noise sources in these heterogeneous nanostructures. Validity of the presented model is verified experimentally for resistors operating under extreme working conditions. For the experimental purposes, thick-film resistors of different sheet resistances and geometries, realized using commercially available thick-film resistor compositions, were subjected to high-voltage pulse (HVP) stressing. The obtained experimental results are qualitatively analysed from microstructure, charge transport mechanism and low-frequency noise aspects. Correlation between resistance and low-frequency noise changes with resistor degradation and failure due to high-voltage pulse stressing is observed.

Keywords: thick-film resistors, low-frequency noise, conducting mechanisms, high-voltage pulse stressing, failure analysis

1. Introduction

Thick-film technology that has been in continuous use for decades, mostly in commercial and specialized electronics, is once again increasing interest. The revival of thick-film technology can be attributed to the increasing application of ceramic micro-electro-mechanical systems (C-MEMSs) and the communications industry's need for electronic circuitry with increased functional capability, reduced weight, improved reliability and environmental stability. When C-MEMS are in question, thick-film technology provides simultaneous realization of sensor and actuator elements as well as electronic circuitry for signal processing. In addition, thick-film resistors, the key assets of thick-film technology, are being used both as sensing and as

INTECH
open science | open minds

© 2017 The Author(s). Licensee InTech. This chapter is distributed under the terms of the Creative Commons Attribution License (http://creativecommons.org/licenses/by/3.0), which permits unrestricted use, distribution, and reproduction in any medium, provided the original work is properly cited.

resistive elements. This new application of thick-film resistive materials leads to reduction in resistor dimensions, higher required tolerances and increasing use of buried components. On the other hand, increasing application of thick-film devices in communication systems requires better knowledge of their modulation effects in these systems that are correlated to low-frequency noise sources in thick resistive films. Since noise investigations are powerful tools in reliability investigations it is of the great importance to determine whether standard low-frequency noise measurements can be used in evaluation of these complex structures. Low-frequency noise in thick-film resistors depends on their microstructure and for that reason it can be used to track structural changes caused by different types of stressing conditions that affect reliability of the film. Relationship between low-frequency noise and structure of thick resistive films has mostly been investigated experimentally. The theoretical problem is not simple because of the thick-film's quite complex microstructure. The variety of the parameters that has to be taken into consideration is certainly one of the most significant limitations. However, data related to thick-film compositions and technological processes can provide information about ranges of parameter values. Moreover, the formation of conducting paths and metal-insulator-metal (MIM) units requires the special attention. For these reasons, this chapter focuses on the correlation between noise parameters and parameters of noise sources in thick-film resistors. Firstly, a model of low-frequency noise in thick resistive films that relate noise parameters to thick-film structural and electrical characteristics is described. Then, failure analysis of thick-film resistors subjected to high-voltage pulse stressing is presented based on resistance and low-frequency noise measurements. At the closing subsection, the brief summary of the topic is presented with an emphasis on the possibility that standard resistance and noise measurements can be used in degradation and failure analysis of thick resistive structures under a wide range of extreme working conditions.

2. Low-frequency noise in thick-film resistive structures

Transport of electrical charges in thick-film materials takes place via chains of conducting particles (**Figure 1**) [1]. Two adjacent conducting particles in the chain can be sintered or insulated by a thin, glass barrier thus determining the electrical current flow. Therefore, metallic conduction and direct tunnelling are dominant conducting mechanisms present in thick resistive films. Tunnelling via traps is also present in thick-film resistors but it is a dominant conducting mechanism when thick insulating layers are in question. Since the insulator layers are thin [2], direct tunnelling is one of the dominant conduction mechanisms [3]. For low applied voltages $V_B \gg \Phi_B$, where Φ_B is the height of the potential barrier, the applied voltage and the current are proportional. Resistance of the resistor is determined by the barrier resistance. Trap at the location x_1, shown in **Figure 1**, is being randomly occupied by electron. Therefore, its presence modulates the direct tunnelling current. Because of the barrier height, field and thermal injection effects are neglected.

Low-frequency noise sources in these noisy devices are correlated to following conducting mechanisms:

http://dx.doi.org/10.5772/intechopen.69442

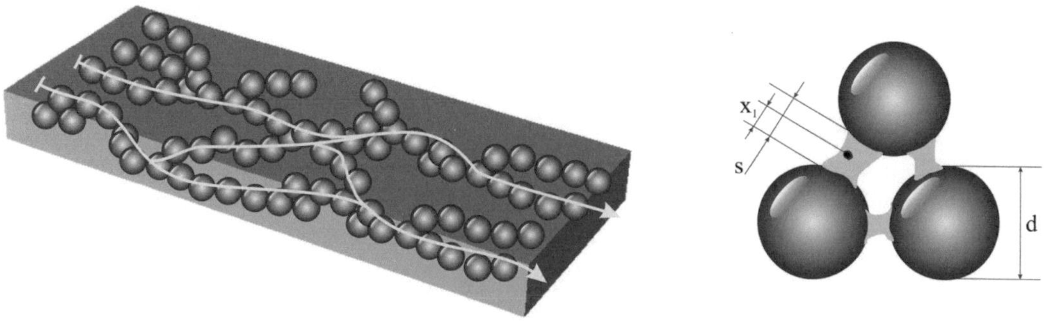

Figure 1. Schematic presentation of thick-film resistor and a segment of a chain where adjacent conducting particles are separated by thin insulating layers.

1. Metallic conduction

Hooge's empirical expression [4] describes low-frequency relative voltage noise spectrum for conduction through conductive grains and contacts between them:

$$\frac{S_{\text{VC}}}{V_C^2} = \frac{\alpha}{V_{\text{ef}} n f}, \tag{1}$$

where α is the Hooge numerical parameter of the order of 10^{-3}, while V_{ef} and n are the effective volume of sintered contact and free carriers concentration in the contact region, respectively. Hooge's expression is used as the empirical relation with the effective parameters in order to express $1/f$ noise, which is a consequence of contact and particle resistivity fluctuations.

2. Tunnelling processes

When conduction through glass barriers is in question, low-frequency fluctuations of tunnelling processes are correlated to the glass matrix space charge fluctuations. These fluctuations can be caused by the presence of traps in glass barriers and fluctuations caused by the thermal noise in the glass matrix [5]. If it is assumed that potential barrier height fluctuations are caused by Nyquist noise of the insulator, then the relative voltage noise spectrum due to the Nyquist noise modulation can be given by the following expression [1, 5]:

$$\frac{S_{\text{VBN}}}{V_B^2} = \frac{4\pi mqkT}{3\Phi_B h^2} \cdot tg\delta \cdot \frac{s^2}{C} \cdot \frac{1}{f}, \tag{2}$$

where q and m are the electron charge and its effective mass, h is the Planck's constant, k is the Boltzmann's constant, T is the absolute temperature, s is the thickness of the insulating layer, $tg\delta$ is the loss tangent of the insulator, Φ_B is the potential barrier height measured with respect to the Fermi energy and C is the capacitance of metal-insulator-metal (MIM) unit [1] that consists of two spherical conducting particles separated by a thin insulating layer.

If it is assumed that MIM insulating layers contain traps, the trap may be of neutral or negative charge. The trap may have negative charge as a consequence of occupation by electron during the tunnelling process. The trap occupation function fluctuation induces the barrier height fluctuation due to the local charge fluctuation. In calculations of noise spectrum due to the trap occupation fluctuations, the following is taken into account:

i. the greatest contribution to the noise is the traps with energies equal to the Fermi level in the conducting particle,

ii. the potential barrier is rectangular and of height Φ_B and width s, and

iii. the applied voltage V_B is low, i.e. $V_B \gg \Phi_B$.

In that case, the relative voltage noise spectrum due to the presence of traps in glass barriers is given by the following expression [6, 7]:

$$\frac{S_{VB}}{V_B^2} = \frac{8\pi m q^3 s^2 x_1^2 \chi}{h^2 A^2 \Phi_B \varepsilon_0^2 \varepsilon_r^2} \frac{\theta}{(1+\theta)^2} \frac{\tau}{1+\omega^2\tau^2}, \tag{3}$$

where x_1 is the position of the trap (**Figure 1**), A is the barrier cross-section, ε_0 is the vacuum electrical permittivity, ε_r is the relative electrical permittivity of the glass, while χ and θ are parameters given by the following expressions:

$$\chi = \left(1 - \frac{x_1}{s}\right)^2 \left[\left(1 - \frac{x_1}{s}\right)^2 + \frac{x_1^2}{s^2}\right], \tag{4}$$

$$\theta = \frac{\tau_c}{\tau_e}. \tag{5}$$

The reciprocal time parameter τ is defined as

$$\tau^{-1} = \tau_c^{-1} + \tau_e^{-1} = C_1 n(E) \exp\left(-2\int_0^{x_1} |k| dx\right) + C_2 \exp\left(-2\int_{x_1}^{s} |k| dx\right), \tag{6}$$

where $|k|$ is the electron wave vector magnitude. The concentration $n(E)$ is the free electron concentration in the conducting particle, with energy E equal to the Fermi energy E_{fm}. C_1 and C_2 are constants of electron capture and emission. The random occupation of the trap depends on the random tunnelling of the electron between two particles separated by a thin glass barrier that contains the trap. As the consequence of the presence of the single-energy level trap, Lorentzian terms may be present in low-frequency noise spectrum of the resistor. The distribution of the trap energy levels can result in noise spectrum with $1/f^{\gamma}$ dependence ($\gamma = 1$ or $\gamma \neq 1$) [8].

http://dx.doi.org/10.5772/intechopen.69442

Under assumption that thick-film resistor can be viewed as the complex network [1] that consists of M parallel chains and that one chain consists of K_C contacts and K_B MIM units, then the total resistance of the resistor is given by

$$R = \frac{K_C R_C + K_B R_B}{M^2}, \tag{7}$$

where R_C is the resistance of the sintered contact between two neighbouring conducting particles and R_B is the barrier resistance. Resistance of the sintered contact between two adjacent conducting particles is determined by the specific resistance of the contact ρ and the radius of the barrier cross section a [1]:

$$R_C = \frac{\rho}{\pi a}. \tag{8}$$

In the case of neighbouring particles separated by a thin glass barrier, resistance of the MIM unit is determined by the barrier resistance [1]:

$$R_B = \frac{h^2 s}{q^2 A (2mq\Phi_B)^{1/2}} \exp\left[\left(\frac{32\pi^2 mqs^2\Phi_B}{h^2}\right)^{1/2}\right]. \tag{9}$$

The tunnelling area A can be calculated as $A = \pi a^2$, where a is the radius of the area that depends on the diameter d of the conducting particle.

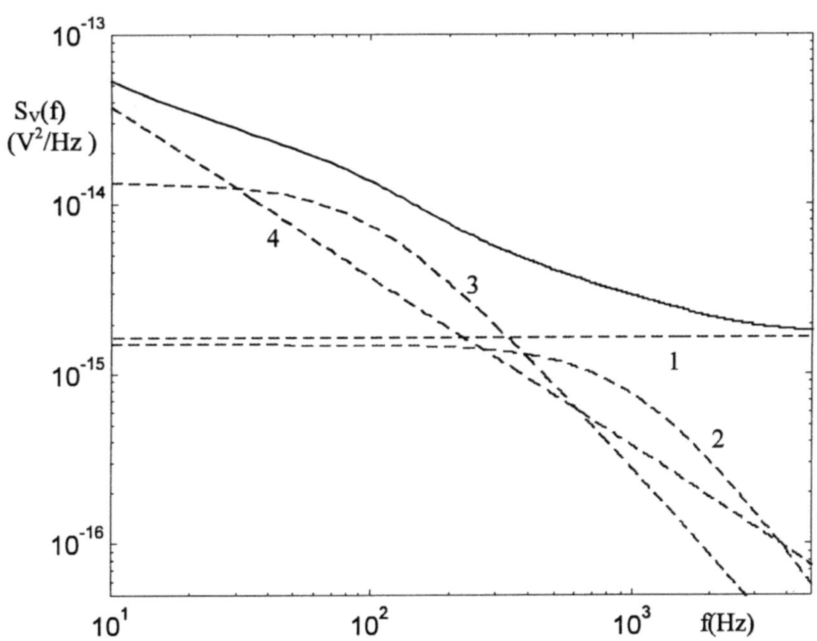

Figure 2. Contributions of different kinds of the noise sources in the total voltage noise spectrum (continuous line) as calculated using Eq. (10) (1—thermal noise; 2, 3—noise due to the presence of traps in glass barriers; and 4—1/*f* noise) [6].

The total relative voltage noise spectrum is therefore given by the following expression [6]:

$$\frac{S_V}{V^2} = \frac{1}{N_C} \frac{1}{\left(1 + \frac{N_B R_B}{N_C R_C}\right)^2} \frac{S_{VC}}{V_C^2} + \frac{1}{N_B} \frac{1}{\left(1 + \frac{N_C R_C}{N_B R_B}\right)^2} \left\{ \frac{S_{VBN}}{V_B^2} + \frac{S_{VB}}{V_B^2} \right\}. \tag{10}$$

where N_B, N_T and N_C are the total number of barriers, traps and contacts, respectively, taking part in the conduction process between two opposite electrodes.

Contributions of different kinds of noise sources included in Eq. (10) are shown in **Figure 2**. These results are obtained by numerical simulation for parameter values: $s = 1.33$ nm, $d = 150$ nm, $a = 6.2$ nm, $N_B = 3.8 \times 10^8$, $N_C = 4.6 \times 10^9$, $N_T = 6.9 \times 10^9$ and $R = 100$ kΩ [6]. $1/f$ noise and noise due to the presence of traps in glass barriers are included along with two Lorentzian terms ($f_{C1} =$ 115 Hz and $f_{C2} = 1.1$ kHz). The contribution of the noise due to the conduction through the conductive grains or contacts between the adjacent ones is negligible, and, therefore, is not shown in **Figure 2**.

3. Failure analysis of thick-film resistors subjected to high-voltage pulse stressing

Different conditions of thick-film resistor application have induced the need to investigate their behaviour under stress, especially high-voltage pulse stress. The most of the published data dealt with trimming of thick resistive films by energy of high-voltage pulses (HVP trimming) [9, 10]—a trimming method based on internal discharges using both thick-film resistor terminations as electrodes for applying the high-voltage energy to the resistor body. Moreover, several papers explored properties of thick-film surge resistors [11] that serve as protection of communication systems from a variety of voltage disturbances such as short duration, high-voltage transients caused by lightning strikes or longer duration over voltages. Nowadays, the revival of thick-film technology that can be attributed to new applications of thick-film resistors induced the necessity of extensive behavioural studies related to undesirable high-voltage pulse stressing of conventional thick resistive films [12–14].

In order to qualitatively analyse the influence of high-voltage pulsing on thick-film resistors, pulse performances have been investigated using a model of low-frequency noise in thick resistive films presented in the previous chapter. Behavioural analysis of thick-film resistors subjected to high-voltage pulse stressing was performed using several groups of thick-film test samples with different resistor geometries (**Figure 3**) realized using commercially available RuO_2 and $Bi_2Ru_2O_7$ mixture-based-thick-film resistor compositions in combination with Pd/Ag conductor composition. Test samples were formed on ceramic alumina (96% Al_2O_3) substrates using conventional screen-printing techniques. After 15 min levelling at 21°C, wet layers were dried at 150°C in a conveyer infrared drier for 10 min. Dry resistive films were 25 ± 3 μm thick. Firing was performed using standard 30-min cycle ($T_{max} = 850$°C, $t_{max} = 10$ min). Pulse performances have been investigated using 100/700 μs pulses delivered by Haefely P6T impulse tester

Figure 3. Thick-film test resistors of different lengths and widths used in experimental investigations.

with an output resistance of 25 Ω. Following experimental set-up was used: 1.5–4.0 kV voltage range, 10 pulse series, 6 pulses/min frequency, $T = 21°C$. Testing conditions were selected in compliance with ITU-T K.20 standard that refers to the resistibility of telecommunication equipment to over-voltages and over-currents. Keithley nanovolt amplifier, Model 103A, was used for voltage noise spectrum measurements along with Hewlett-Packard dynamic signal analyzer 3561B in the 10 Hz to 10 kHz frequency range. A noise index [15] was also measured in compliance with the test method standard for electronic and electrical component parts MIL-STD-202D, Method 308, at 1 kHz. For noise index measurements, Quan-Tech Resistor Noise Test Set, Model 315B was used. Hewlett-Packard 34401A instrument was used for resistance measurements.

Table 1 and **Figure 4** [12] present typical results obtained by noise and resistance measurements for 10 and 100 kΩ/sq test resistors that were exposed to the high-voltage treatment. Resistors with identical 1 × 2 mm^2 geometries suffered degradation but they did not catastrophically fail. Results obtained by noise index, voltage noise spectrum and resistance

Degradation			**Catastrophic failure**		
R_{sq} (kΩ/sq)	10	100	R_{sq}(kΩ/sq)	10	100
R_N (kΩ)	16	220	R_N (kΩ)	8	110
$\overline{R_i}$ (kΩ)	15.942	220.060	R_i (kΩ)	7.842	111.92
$\overline{R_s}$ (kΩ)	15.490	217.650	R_S (kΩ)	22.55	11.8(→105.5)
$\overline{S_{R_i}}$ (Ω^2/Hz)	2.508×10^{-7}	2.45×10^{-4}	S_{R_i}(Ω^2/Hz)	9.89×10^{-8}	2.48×10^{-4}
$\overline{S_{R_s}}$ (Ω^2/Hz)	5.2289×10^{-5}	2.4×10^{-3}	S_{R_s} (Ω^2/Hz)	1.5×10^{-2}	7.13×10^{-4}
$\overline{NI_i}$ (dB)	−23.8	−7.5	NI_i (dB)	−15.2	7
$\overline{NI_s}$ (dB)	−0.8	0.5	NI_s (dB)	21.7	24.2

Table 1. Resistor parameters before (i) and after (s) high-voltage pulse stressing [12].

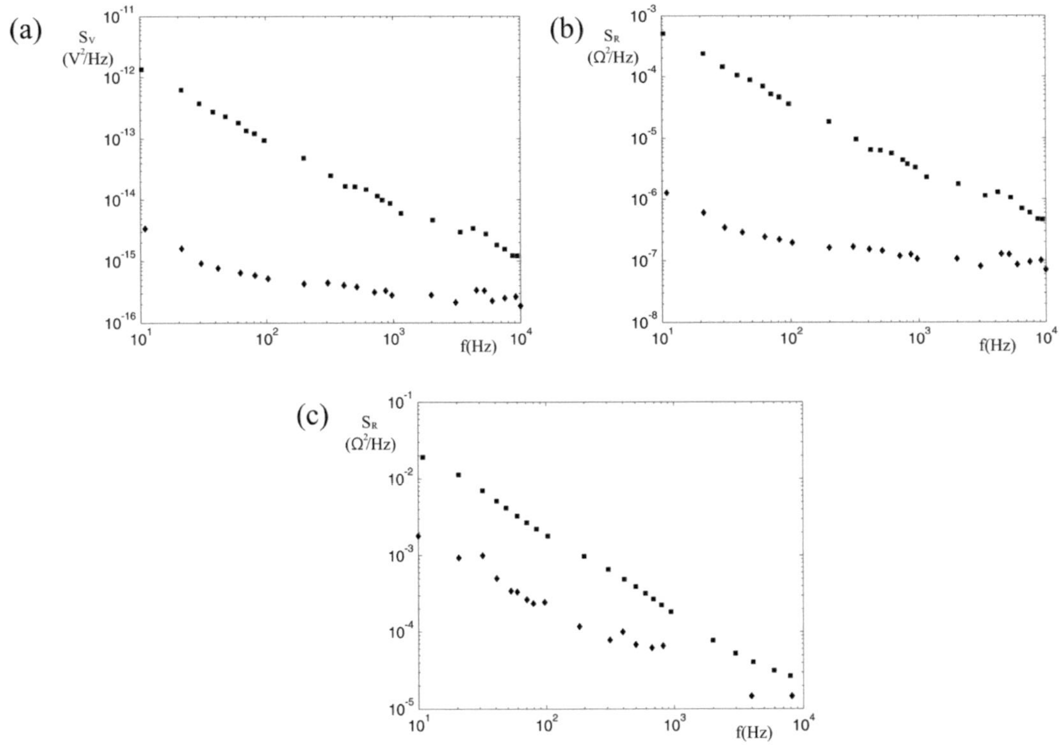

Figure 4. Experimental results for voltage and resistance noise spectra (◆—before pulse stressing, ■—after pulse stressing), for thick-film resistors with following nominal resistances: $R = 16$ kΩ (a, b) and $R = 220$ kΩ (c) [12].

measurements are given for resistors with a nominal resistance of 16 kΩ that were subjected to the impact of eleven 1500 V pulses and a single 3000 V pulse and resistors with a nominal resistance of 220 kΩ that were subjected to the impact of eleven 1500 V pulses and eleven 3000 V pulses. After impacts of the first and the tenth pulse from the series of pulses with the same amplitude, voltage noise spectrum and resistance measurements were performed. Results for two series of degraded resistors (10 and 100 kΩ/sq) and two resistors that suffered catastrophic failure are given in **Table 1**. Sheet resistances, R_{sq}, nominal resistances, R_N, resistances, R, resistance noise spectra, S_R, and noise indexes, NI, before (i) and after (s) high-voltage pulse treatment are presented.

Results given in **Table 1** show that resistance decreases with high-voltage pulse treatment of thick resistive films. The relative resistance change for both groups of resistors is of the order of several percents. This change although small is higher for 10 kΩ/sq resistors (3%) than for 100 kΩ/sq resistors (1%).

Figure 4 presents the experimental results for resistance noise spectra before and after high-voltage pulse stressing. Since the voltage noise spectrum depends on current I, the measured values for the voltage noise spectrum were used for resistance noise spectrum ($S_R = S_V/I^2$) calculations. Compliance of voltage and resistance noise spectra, which is in agreement with

Ohms law, is shown in **Figure 4a** and **b**. An increase of resistance noise due to high-voltage treatment is observed. Resistors based on compositions with higher sheet resistances exhibit smaller final change of the resistance noise. Since the same test conditions were applied to all test samples, the obtained results were expected. Results for all test resistors presented in **Table 1** show that measured resistance changes are less distinguished than the measured resistor noise spectrum changes.

High-voltage pulse stressing caused microscopic changes in thick resistive films that manifested in presented results. Thick resistive films are complex conductive networks. These conductive networks are result of the sintering processes. Transport of electrical charges takes place via a number of conducting chains. These chains consist of clusters of particles (particles that are in contact) and neighbouring particles separated by thin glass barriers (MIM units). Therefore, the current flow is being determined by metallic conduction through clusters of particles and tunnelling through glass barriers. Multiple tunnelling processes take place when the traps are present in glass barriers. Impurities introduced during technological processes and partial dissolution of metal-oxide in glass are responsible for the presence of traps. During high-voltage treatment, resistance change occurs due to barrier and contact resistance changes. High-voltage pulse stressing induces electrical field inside metal-insulator-metal unit that is not sufficient to induce dielectric breakthrough and therefore a decrease in the resistance due to the increase in a number of contacts between neighbouring particles does not occur. It is more likely that high-voltage treatment affects electrical charges captured within thin glass layers between neighbouring conducting particles or that the concentration of traps increases due to changes in microstructure of the resistor thus affecting noise performances of the resistor more than resistance values. Besides that, resistance decrease may occur due to the conversion of single chain from the non-conducting state to the conducting state. Under the same straining conditions, depending on the volume fraction of the conductive phase, thick-film resistor exhibits different changes in resistance values. A conductive/insulating phase ratio determines the microstructure of the thick resistive film and present conducting mechanisms as it is shown in scanning electron microscopy (SEM) micrographs of 10 and 100 kΩ/sq thick-film resistors given in **Figure 5**. It can be seen that resistors based on compositions with greater sheet resistances have greater content of the glass phase. For that reason the most of the neighbouring conducting particles are separated by thin glass barriers. In that case, the conducting mechanism known as multiple tunnelling becomes dominant. On the other hand, resistors based on compositions with lower sheet resistances have lower content of the glass phase and therefore large conductive areas are present. In that case, the conducting mechanism known as metallic conduction is also present. This also means that for the same voltage pulse, a greater electric field can be achieved within the thin glass layer between two neighbouring particles along with the greater current. For these reasons, resistors with smaller sheet resistances exhibit greater resistance changes caused by high-voltage treatment.

Measurements of the noise voltage spectrum showed that high-voltage treatment results in the increase of noise voltage and corresponding resistance noise spectra. Moreover, dominant contribution of the $1/f$ noise source is observed. High-voltage pulse stressing affects electrons captured by traps in thin glass barriers that are not directly involved in the conduction. However, conduction is being modulated by electrical charges captured by traps that alter the height of the potential barrier of metal-insulator-metal unit. For these reasons, change in the resistance noise

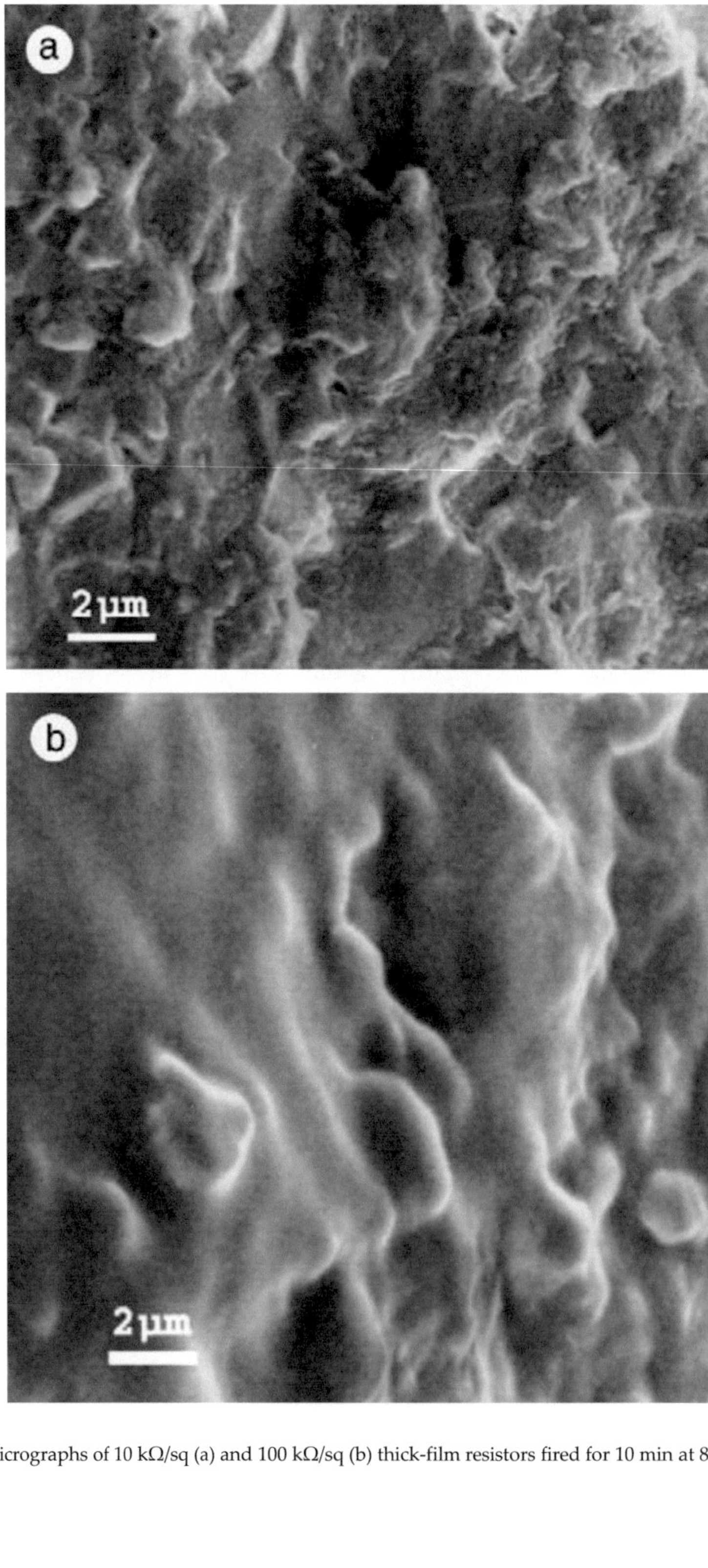

Figure 5. SEM micrographs of 10 kΩ/sq (a) and 100 kΩ/sq (b) thick-film resistors fired for 10 min at 850°C [12].

http://dx.doi.org/10.5772/intechopen.69442

spectrum is considerably higher than change in resistance values before and after high-voltage pulse stressing. It confirms the presumption that low-frequency noise exhibits greater sensitivity to changes at the microstructural level than resistance. Besides that, resistors formed using compositions with higher sheet resistances exhibit smaller resistance noise spectrum changes. Since it is noted that contacts between neighbouring conducting particles negligibly affect low-frequency noise, such behaviour can be attributed to differing spatial distributions of traps in resistors based on compositions with different sheet resistances. Along with measurements of noise voltage spectrum, noise index [15] measurements were performed since noise index is well known as one of the standard quality and reliability indicators used in the fabrication of thick resistive films. Noise index was measured before and after performed stressing. Measurements showed increasing noise index values due to applied stressing. As expected, the observed noise index increase is in agreement with the resistance noise spectrum measurements having in mind the noise index definition. However, noise spectra measurements can provide additional information regarding noise nature and sources that are related to microstructural properties of thick resistive films and present conducting mechanisms.

High-voltage treatment caused numerous catastrophic failures in tested resistors [12]. **Figure 6** shows failed 100 kΩ/sq thick-film resistor that was exposed to eleven 1500 V pulses and a single 3000 V pulse. A resistor with the initial resistance of $R = 111.92$ kΩ was covered with fine Ag powder that migrated from the contact area between resistive film and conducting path. Ag powder affected resistance of the resistor by decreasing its value. After the mechanical removal of the powder resistor regained its 100 kΩ value and damaged area became visible. As it can be seen from **Figure 6a**, the segregation of the resistive film took place. The cause of the failure is probably a faulty technological process that introduced impurities in the resistive film. The micrograph of the failure spot shown in **Figure 6b** showed that thin resistive layers still remained present in the failure region. The ratio between the glass and

(a)
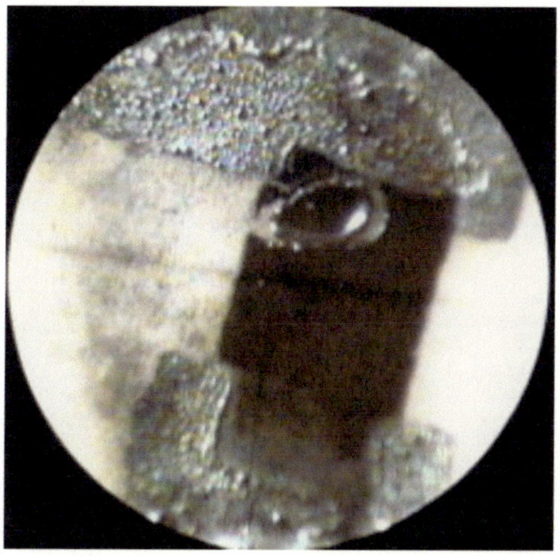

(b)
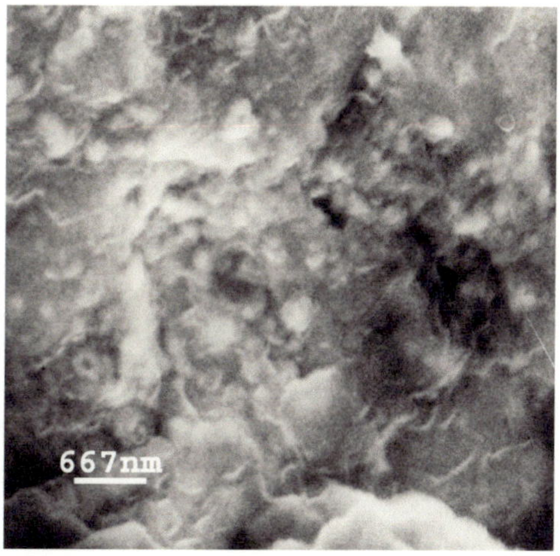

Figure 6. Photograph of 100 kΩ/sq thick-film resistor that catastrophically failed (a) and micrograph of the damaged area (b) [12].

conductive phase was changed because of the higher concentration of the conductive phase in the surface layer of the resistor. Several factors affected the performances of this resistor. High-voltage treatment affected both the microstructure and macrostructure of the resistor, introduced visible physical damage and damage caused by mechanical removal of thin layer of Ag powder.

In order to prove that high-voltage treatment caused microstructural changes in noise index and resistance noise spectra were measured. **Figure 7** and **Table 1** show that noise performances of the failed resistor were in correlation with noise performances of resistors that did not suffer failure. The segregated area accidentally did not strongly affect microstructure of the resistor. A thin conducting layer that remained at the failure spot probably had a shunting effect that compensated decreased thickness of the resistive film.

Figure 8a shows the failed 10 kΩ/sq thick-film resistor. The conducting film, as well as the contact area between resistive and conductive film, was damaged after the impact of a single 1500 V pulse. The possible cause of this occurrence may be defect migration or poorly formed contact between resistor and neighbouring conducting path. Resistance increase confirmed that the resistor area was affected. However, it cannot be concluded how this defect influenced the

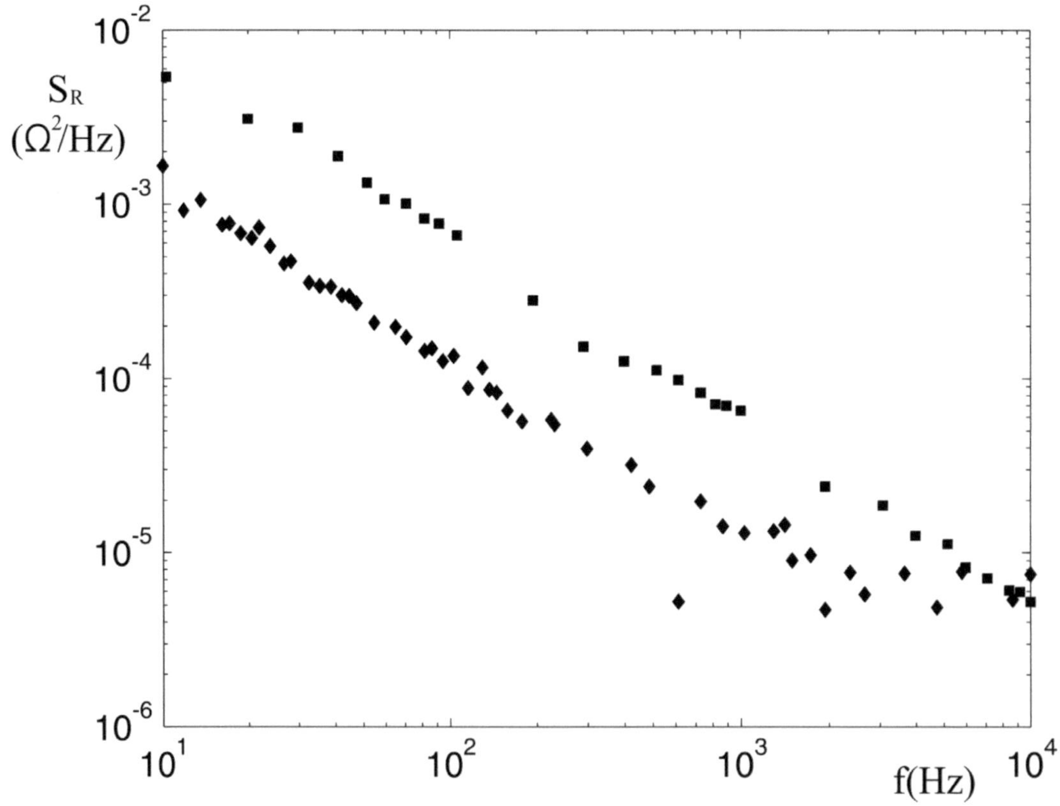

Figure 7. Experimental results for resistance noise spectrum (◆—before pulse stressing, ■—after pulse stressing) for catastrophically failed thick-film resistor with the initial resistance $R = 111.92$ kΩ [12].

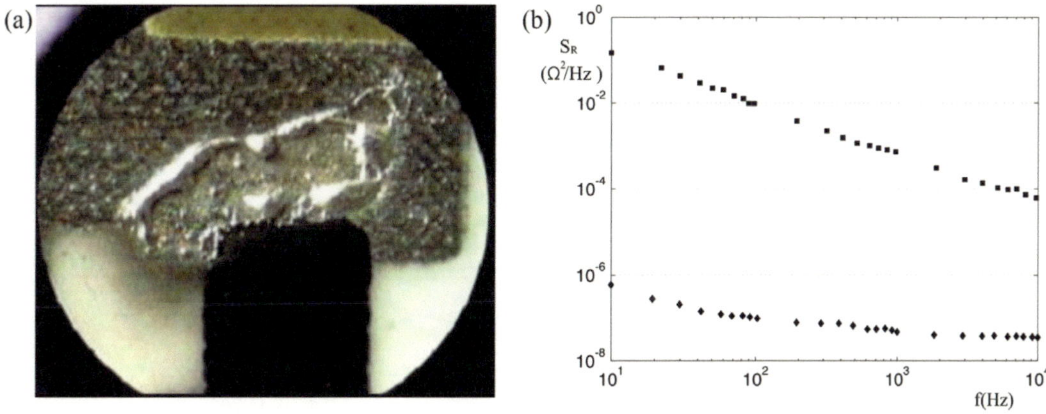

Figure 8. Photograph of catastrophically failed thick-film resistor (a) with the initial resistance $R = 7.842$ kΩ and experimental results for resistance noise spectrum (◆—before pulse stressing, ■—after pulse stressing) (b) [12].

frequency-dependent part of the low-frequency noise spectrum. According to the results given in previous figures, the resistance noise spectrum and noise index increases (**Figure 8b** and **Table 1**) are probably mainly related to changes in the microstructure of the resistive layer.

Figure 9 shows catastrophically failed resistor with a sheet resistance of 1 kΩ/sq [14]. Resistance of the thick-film resistor gradually increased with high-voltage pulse treatment until pulse amplitude reached its critical level at which resistor suffered catastrophic failure due to the excess loaded voltage. Both resistive film and conducting path were visibly damaged. As expected, pulse stressing also affected noise performances of the resistor. Noise index gradually increased with applied stressing until failures occurred resulting in maximal noise index values (**Figure 9a**). **Figure 9b** illustrates this mode of catastrophic failure.

During the high-voltage pulse treatment, destruction of the resistor may also occur. Conducting path degradation may lead to the dispersion of the conductive material to the resistor area resulting in the presence of local hot spots and resistor burning and evaporation. A typical example of this mode of catastrophic failure is shown in **Figure 10**.

During testing, several encapsulated resistors suffered progressive resistor degradation that led to catastrophic failure [13]. The photograph of the characteristic mode of progressive 10 kΩ/sq resistor degradation due to thermal effects induced by high-voltage treatment is given in **Figure 11**. Note that 10 kΩ/sq resistor was subjected to high-voltage pulse treatment using pulses with 3 and 4 kV amplitudes. With each applied 3 kV pulse resistor gradually degraded. At first, glass encapsulant started to melt and with further stressing several areas of thick resistive film became fully exposed to the environment. After increasing the pulse amplitude to 4 kV catastrophic failure took place. Thick resistive film burned and evaporated. The volume of the resistor decreased resulting in 430% resistance increase. The reported noise index values were in accordance with resistance values. Unacceptably high resistance and noise index values registering strain induced degradations along with diminished integrity of the resistive layer qualified this test resistor as unreliable.

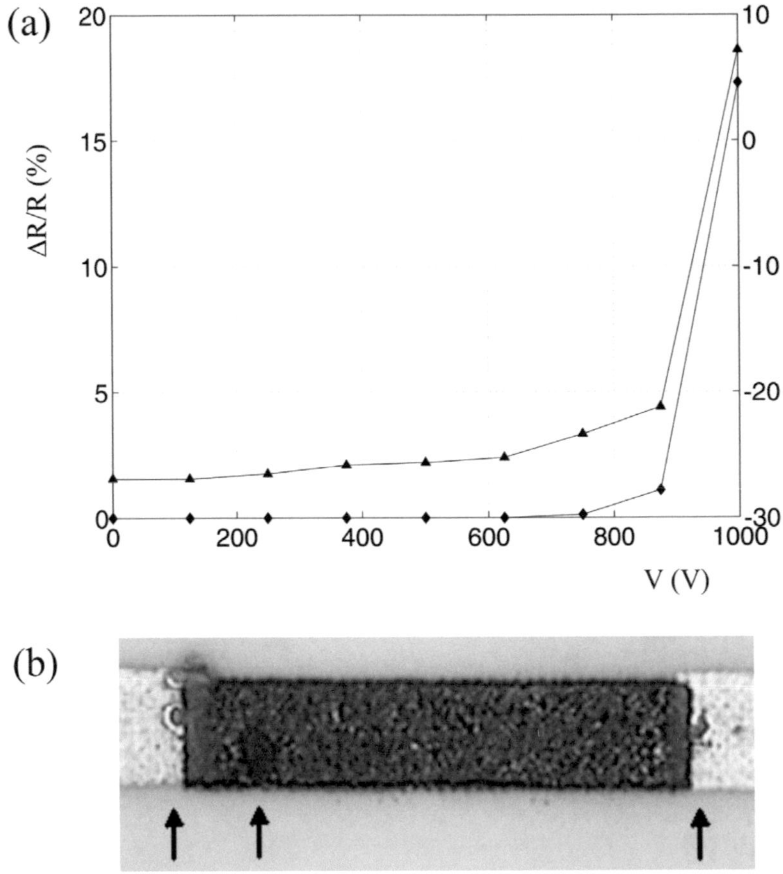

Figure 9. Experimental results for relative resistance changes and noise index during high-voltage pulse stressing of 1 kΩ/sq thick-film resistor (a) and a photograph of catastrophically failed resistor with designated failure points (b) [14].

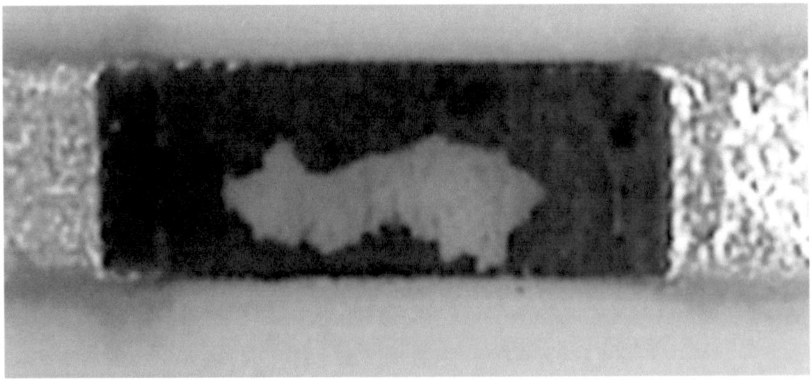

Figure 10. Photograph of catastrophically failed thick-film resistor due to the presence of hot spots caused by conducting material dispersion.

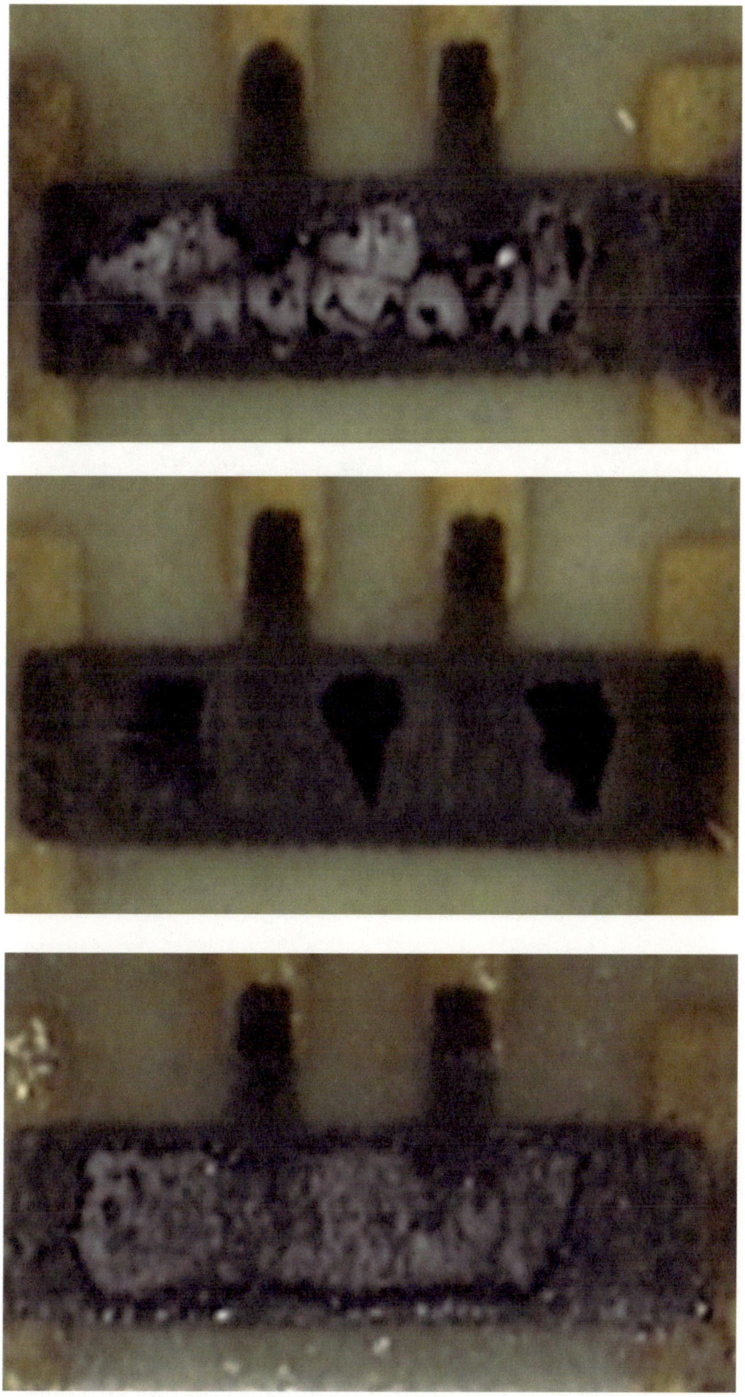

Figure 11. Photographs of progressive resistor degradation due to high-voltage pulse stressing: melting of glass encapsulant, direct exposure of resistive layer to surrounding atmosphere and burned and partially evaporated resistive layer [13].

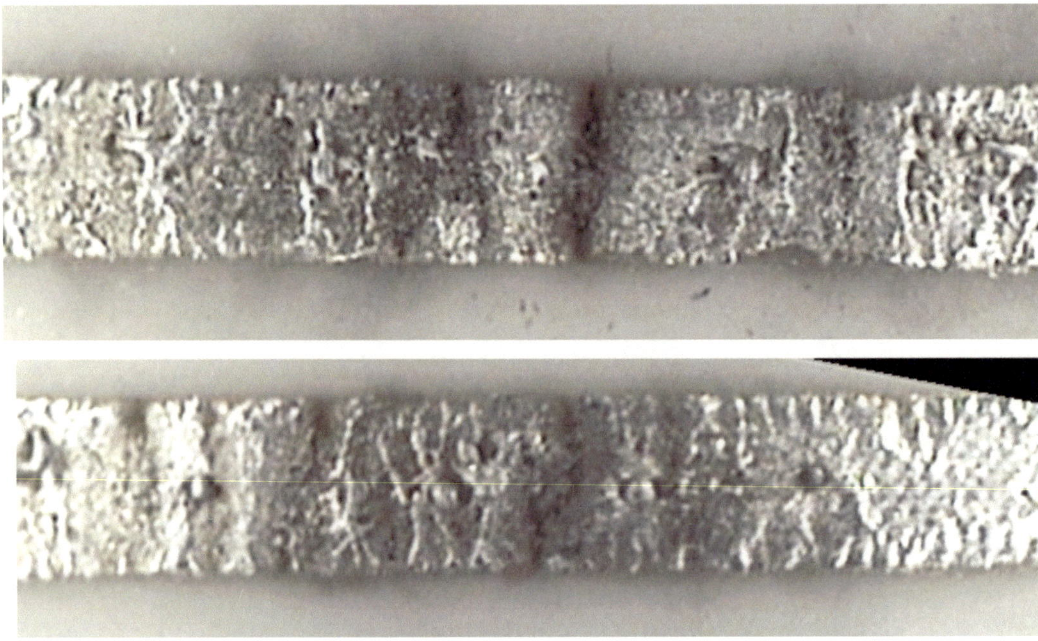

Figure 12. Photographs of failed conducting paths.

It should be mentioned that the failure of resistive layers due to high-voltage pulse stressing is often accompanied by the failure of conducting paths. The characteristics of failure modes for thick-film conducting paths are shown in **Figure 12**. The high-voltage treatment may diminish conducting path integrity causing burning and evaporation of conducting path segments.

4. Conclusion

Degradation and failure analysis of thick-film resistors is identified as the constant manufacturers challenge due to the growing market of C-MEMS devices and reliable communication systems. These contemporary applications of thick resistive materials induced the need to investigate their behaviour under various stressing conditions, especially electrical stressing conditions. On the other hand, there is a growing interest in noise measurements as means of thick-film resistor quality evaluation and evaluation of degradation under stress. For these reasons, this chapter presented the study of high-voltage pulse stressing effects on thick-film resistors based on the model of low-frequency noise in thick-film structures based on close relationship of the noise and conduction mechanisms. Correlation between resistance and low-frequency noise changes and high-voltage pulse stressing was observed and qualitative degradation and failure analysis was performed based on standard noise and resistance measurements. Several catastrophically failed resistors were presented and their failure modes were analysed. Results presented in this chapter confirmed that standard resistance and noise measurements can be used in degradation and failure analysis of thick resistive structures.

They aim to open new prospects for further investigations and quantitative analysis that may result in a method of diagnostic of microstructure effects as well as improved quality assessment of thick-film resistors under a wide range of extreme working conditions.

Acknowledgements

The authors would like to thank the Ministry of Education, Science and Technological Development of the Republic of Serbia for supporting this research within projects III44003 and III45007.

Author details

Ivanka Stanimirović

Address all correspondence to: inam@iritel.com

Institute for Telecommunications and Electronics IRITEL a.d. Beograd, Belgrade, Republic of Serbia

References

[1] Kusy A, Szpytma A. On 1/f noise in RuO_2-based thick resistive films. Solid-State Electronics. 1986;**29**:657-665. DOI: 10.1016/0038-1101(86)90148-6

[2] Chiang Y-M, Silverman LA, French RH, Cannon RM. Thin glass film between ultrafine conductor particles in thick-film resistors. The Journal of the American Ceramic Society. 1994;**77**:1143-1152. DOI: 10.1111/j.1151-2916.1994.tb05386.x

[3] Pike GE, Seager SH. Electrical properties and conduction mechanisms of Ru-based thick-film (cermet) resistors. Journal of Applied Physics. 1977;**48**:5152-5169. DOI: 10.1063/1.323595

[4] Hooge FN. 1/f noise is no surface effect. Physics Letters A. 1969;**29A**:139-140. DOI: 10.1016/0375-9601(69)90076-0

[5] Kleipenning TGM. On low-frequency noise in tunnel junctions. Solid-State Electronics. 1982;**25**:78-79. DOI: 10.1016/0038-1101(82)90100-9

[6] Mrak I, Jevtić MM, Stanimirović Z. Low-frequency noise in thick-film structures caused by traps in glass barriers. Microelectronics Reliability. 1998;**38**:1569-1576. DOI: 10.1016/S0026-2714(98)00032-8

[7] Jevtić MM, Stanimirović Z, Mrak I. Low-frequency noise in thick-film resistors due to two-step tunnelling process in insulator layer of elemental MIM cell. IEEE Transactions on Components, Packaging, and Manufacturing Technology. 1999;**22**(01):120-127. DOI: 10.1109/6144.759361

[8] Pellegrini B. $1/f^{\gamma}$ noise from single-energy-level defects. Physical Review B. 1987;**35**(2): 571-580. DOI: 10.1103/PhysRevB.35.571

[9] Ehrhardt W, Thrust H. Trimming of thick-film-resistors by energy of high voltage pulses and its influence on microstructure. In: Proceedings of 13th European Microelectronics and Packaging Conference, May 31st–June 1st, Strasbourg, France, 2001; 403-407

[10] Dziedzic A, Kolek A, Ehrhardt W, Thust H: Advanced electrical and stability characterization of untrimmed and variously trimmed thick-film and LTCC resistors. Microelectronics Reliability. 2006;**46**:352-359. DOI:10.1016/j.microrel.2004.12.014

[11] Barker MF. Low Ohm resistor series for optimum performance in high voltage surge applications. Microelectronics International. 1997;**43**:22-26. DOI: 10.1108/13565369710800493

[12] Stanimirović I, Jevtić MM, Stanimirović Z. High-voltage pulse stressing of thick-film resistors and noise. Microelectronics Reliability. 2003;**43**:905-911. DOI: 10.1016/S0026-2714(03)00094-5

[13] Stanimirović I, Jevtić MM, Stanimirović Z. Multiple high-voltage pulse stressing of conventional thick-film resistors. Microelectronics Reliability. 2007;**47**:2242-2248. DOI: 10.1016/j.microrel.2006.11.017

[14] Stanimirović Z, Jevtić MM, Stanimirović I. Simultaneous mechanical and electrical straining of conventional thick-film resistors. Microelectronics Reliability. 2008;**48**:59-67. DOI:10.1016/j.microrel.2006.09.039

[15] Jevtić MM, Stanimirović Z, Stanimirović I. Evaluation of thick-film resistor structural parameters based on noise index measurements. Microelectronics Reliability. 2001;**41**:59-66. DOI: 10.1016/S0026-2714(00)00207-9

Chapter 6

Failure Concepts in Fiber Reinforced Plastics

Roselita Fragoudakis

Additional information is available at the end of the chapter

http://dx.doi.org/10.5772/intechopen.71822

Abstract

The anisotropic nature of composite materials, specifically fiber reinforced plastics (FRPs), constitutes them a material category with adaptable mechanical properties, appropriate for the application they are being designed for. The stacking sequence choice of FRP laminates allows for the optimization of their strength, stiffness, and weight to the desired design requirements. The anisotropic nature of composites is also responsible for the different failure modes that they experience, which are based on the accumulation of damage, rather than crack initiation and propagation as the majority of homogeneous isotropic materials. This chapter discusses the background theory for determining the stress distribution in a laminated FRP, the possible failure modes occurring in composites, the failure criteria predicting the onset of failure, as well as cumulative damage models predicting the fatigue life of laminates.

Keywords: anisotropy, transverse isotropy, stacking sequence, fiber reinforced plastics (FRPs), laminates, classical lamination theory, failure criteria, cumulative damage models, fatigue

1. Introduction

Lighter and more durable structures have become the focus of the majority of industries. From aviation and automotive to the battery industry, research and development sectors turn to polymer based reinforced composite materials as an alternative to metals. Fiber reinforced plastics (FRPs) are synthetic composites of epoxy resin and fibrous high strength materials. They have high strength and stiffness, are much lighter than their metal counterparts, and in the majority of cases show high resistance to corrosion [1].

FRPs offer the option of building components tailored to the desired properties for the destined applications. It is specifically FRP laminates, having unidirectional, long fiber reinforcement that can be optimized to provide the desired strength and stiffness at a low desirable

INTECH
open science | open minds

© 2017 The Author(s). Licensee InTech. This chapter is distributed under the terms of the Creative Commons Attribution License (http://creativecommons.org/licenses/by/3.0), which permits unrestricted use, distribution, and reproduction in any medium, provided the original work is properly cited. (cc) BY

weight. As it will be shown later in this chapter, optimizing the fiber orientation in FRPs helps design and build optimum laminated structures.

Composite materials, by definition, are made of two or more constituent materials insoluble in each other. As a result, they are heterogeneous in nature. Specifically FRPs, composed of an epoxy resin matrix and a fibrous reinforcing phase, are heterogeneous and also highly anisotropic. It is imperative therefore, to account for this anisotropic nature and how it can affect the failure mechanisms of FRPs.

However, before the failure concepts in FRPs are addressed, it is imperative to define failure. What is failure, and how is it perceived? Failure is best defined as the point where a component seizes to perform adequately for the application it is designed for. Whether failure is expressed as catastrophic, similar to fracture of a metal component, or as degradation of mechanical properties of the material, for example due to creep, it is important to understand the mechanisms that can lead to any type of undesired failure and design the component against it.

This chapter will address the anisotropic and heterogeneous nature of FRPs, when it is most important to be accounted for, and how it is linked to estimating the strength and mechanical properties of FRPs. It will then discuss the different possible mechanical failure modes in composites and introduce some common criteria to predict the onset of failure. Cumulative damage is a failure concept for FRPs, and it is closely related to fatigue. The effect of some cumulative damage models will be shown.

2. Anisotropic heterogeneous materials

As mentioned above, composite materials are the result of the combination of two or more material phases insoluble in each other. FRPs in particular, have two phases: an epoxy resin matrix and a fiber-reinforcing phase. Each phase offers different properties and qualities to the composite, and depending on how the two phases are combined together, i.e. what volume percent of the total material is occupied by the matrix and the fibers, the properties of the composite will be different.

The matrix phase in FRPs is assumed to be a homogeneous isotropic material. However, it is the reinforcing phase that is responsible for the anisotropic nature of the composite. Fiber reinforcement may take different forms. FRPs may be reinforced by long unidirectional fibers, by long fibers woven in different ways, or short, chopped fibers scattered in the matrix. Although the failure modes discussed later in this chapter refer to all three types of FRPs presented above, the failure criteria and some examples of glass fiber reinforced plastics (GFRPs) discussed below, only concern laminated unidirectional FRPs reinforced with long fibers.

A convenient manufacturing process for FRPs is lamination. Laminated FRPs are structures composed of more than one layer of FRPs, often referred to as plies or laminae. Each lamina has a specified fiber orientation, as well as a given volumetric fraction of each phase. The way in which the laminae of different orientations are stacked on top of each other constitutes the stacking sequence of the laminate, which determines to a large degree the mechanical properties of the laminate.

Many times in the analysis of composite components, an FRP laminate is viewed as a homogeneous, isotropic material having bulk mechanical properties. This is a practice often followed when the laminate is viewed macroscopically and attention is paid more on its geometry and loading conditions, rather than the specific effect of its properties. For example, a laminated beam may be considered as a homogenous component if its deflection is to be estimated under prescribed loading conditions, as is the case of an airplane wing. On the other hand, when investigating the strength, stiffness, and designing against failure, the anisotropic and heterogeneous nature of the FRP should be accounted for. This is when analysis moves from the macroscopic lever to the lamina or even to the microscopic level, investigating each of the constituent materials, matrix and reinforcement, and their interface individually. The failure criteria presented below will focus on the lamina level.

As mentioned above, the composite material mechanical properties are different from those of the constituent materials, and depend on the volume fraction of each phase present. Rules of mixtures, is a simple model that allows calculation of the elastic properties of the FRP, from those of its constituent materials. Rules of mixtures and the Halpin-Tsai equations are shown below as a means of determining Young's modulus (E_i) and Poisson's ratios (v_{ij}), as bulk material properties. The Young's modulus of the composite will vary in the three directions of the material, due to its anisotropic nature, as shown in Eqs. (1) and (2). The Shear Modulus (G_{ij}) (Eq. (3)) is estimated by Halpin-Tsai equations, while the shear moduli G_{23}=G_{32} are estimated using Rules of Mixtures. The bulk moduli (K) are expressed in in Eq. (5 a-c), and are used in calculating Poisson's Ratios using Rules of Mixtures (Eqs. (6) and (7)) [2].

$$E_1 = (1-f)E_m + fE_f \tag{1}$$

$$E_2 = E_3 = E_m \frac{(1+\xi\eta f)}{(1-\eta f)} \tag{2}$$

$$G_{12} = G_{21} = G_{13} = G_{31} = \frac{G_m(1+\xi\eta f)}{(1-\eta f)} \tag{3}$$

where the subscripts m and f, refer to matrix and fiber, respectively, and 1,2,3, to the directionality of the material. The constant f is the volume fraction of fibers in the composite such that $0 \le f \le 1$, $\xi \approx 1$, and

$$\eta = \frac{\left(\frac{E_f}{E_m} - 1\right)}{\left(\frac{E_f}{E_m} + \xi\right)} \text{ or } \eta = \frac{\left(\frac{G_f}{G_m} - 1\right)}{\left(\frac{G_f}{G_m} + \xi\right)} \tag{4}$$

$$K = \left[\frac{f}{K_f} + \frac{(1-f)}{\mathrm{K}_m}\right]^{-1} \tag{5a}$$

$$K_f = \frac{E_f}{3(1 - 2\nu_f)} \tag{5b}$$

$$K_m = \frac{E_m}{3(1 - 2\nu_m)} \tag{5c}$$

$$\nu_{12} = (1 - f)\nu_m + f\nu_f \tag{6}$$

$$\nu_{23} = 1 - \nu_{21} - \frac{E_2}{3K} \tag{7}$$

The above equations show the effect of the constituent materials volume fractions to the stiffness of the composite.

The strength of the FRP becomes not only an important aspect in designing against failure, but also a parameter that can be optimized in order to build a lamina and consequently a laminate of high strength, desirable stiffness, and low weight. As will be shown below, the strength of the FRP can be optimized using an optimum stacking sequence, which involves laminae of different fiber orientations stacked in a specific order. Determining the strength involves therefore, not only accounting for the heterogeneity of the material as in rules of mixtures, but most importantly for the anisotropy of the material, and using this anisotropic nature to affect the properties of the FRP.

Setting up constitutive relationships for FRPs, using the generalized Hooke's Law (Eq. (8)), would require a total of 81 elastic constants in order to fully characterize the material behavior, which could be brought down to 36 constants in the case that both stresses and strains are assumed to be symmetric.

$$\sigma_{ij} = E_{ijkl}\varepsilon_{kl} \tag{8}$$

The nature of FRPs allows for manipulating their anisotropy in the lamina level to further decrease the number of elastic constants required to build constitutive relationships. At the lamina level there are two sets of axis that can be used to express the directionality of the material. One set, the global axis, refers to a reference frame of the laminate, where typically the horizontal, transverse, and vertical directions coincide with the dimensional directions of the laminate. However, each lamina, will have fibers oriented a certain way, therefore a second set of axis, referred to as local or principal axis are used, where the longitudinal direction is always parallel to the longitudinal fibers, thus making an angle with the global horizontal direction equal to that of the orientation angle of the fibers. In this manner each lamina has three mutually orthogonal axis of rotational symmetry, which reduced the number of required elastic constants to 12, where 9 of these are independent. Taking a closer look to the case of a unidirectional FRP lamina it can be observed that the lamina has two axis of symmetry (2,3) making a plane of isotropy in the material (**Figure 1**). This is because the properties of unidirectional FRPs are the same along the 2 and 3 directions, thus plane 23 becomes the isotropy plane. This becomes a special case of orthotropic materials, called transversely isotropic materials, which allows for further decrease of the independent elastic constants from 9 to 5.

The Classical Lamination Theory (CLT), applicable only to orthotropic continuous laminated composite materials, is the set of equations that allows for the development of a constitutive relationship that determines the state of stress in each layer of a laminate [3–5].

CLT investigates lamina, first as a separately structure and then by taking into account its position within the laminate, in order to determine the stress and strain distribution in the laminate. It can therefore, be understood how stacking sequence selection is significant to the strength and performance of a laminate. Consequently, the position of each laminae in the laminate should be clearly defined. There is a specific way to number the lamina in a laminate, and following this configuration the position of the laminae is used in the CLT. Laminae numbering begins from the bottom lamina in a laminate as shown in **Figure 2**. The fictitious separation plane that goes through the mid-section of the laminate, called the mid-surface plane, serves as a datum from where the position of each lamina is determined. As a result, the position of the lamina may be positive, if the lamina is above the mid-surface plane, and negative, when it is below this plane. Such laminae numbering configuration is important and useful in understanding stacking sequence nomenclature, as well as in communicating stress concentration regions in the laminate.

Apart from elastic properties, which can be determined either experimentally or though the Rules of Mixtures and Halpin-Tsai equations, CLT requires knowledge of thermal expansion properties, estimated at each of the three directions of the FRP composite, and in many cases hygroscopic coefficients. In the case of the transversely isotropic materials, only two sets of material properties are required: one set for direction 1 and one for direction 2, which has the

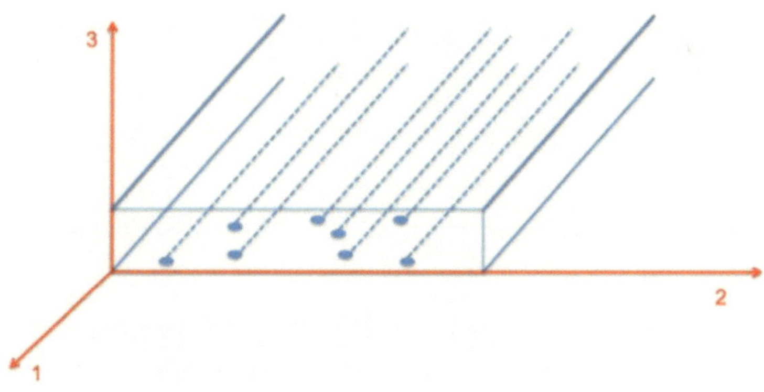

Figure 1. Orientation of the fibers of a unidirectional composite along direction 1.

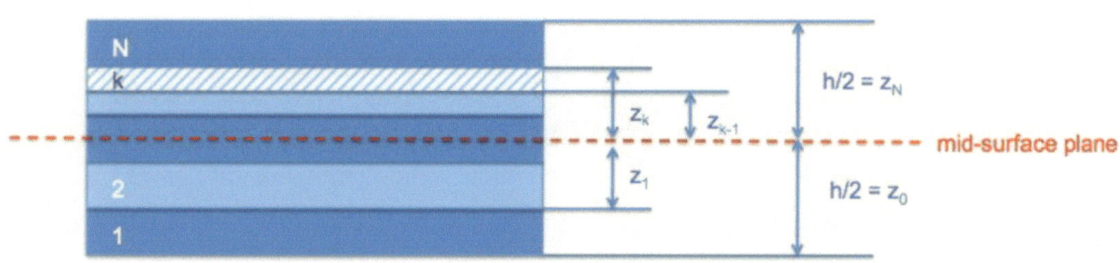

Figure 2. Nomenclature of laminae stacking.

same properties as direction 3. The first step of CLT is to determine stiffness matrices for each lamina. The stiffness matrices (Q and $\overline{Q}$) are evaluated for each lamina and as a result, the effect of the fiber orientation is taken into account (Eq. (9)). The bar over Q shows that the fiber orientation in the matrix is other than 0°, constituting the lamina an off-axis lamina, meaning that it is not along the x-direction of the laminate.[1] The equations for the stiffness matrix element involve elastic property information, as well as transformation matrix components to apply the effects of fiber orientation. The subscript k in Eq. (9) denotes the kth lamina in the laminate.

$$\overline{Q}_k = \begin{bmatrix} \overline{Q}_{11} & \overline{Q}_{12} & \overline{Q}_{16} \\ \overline{Q}_{21} & \overline{Q}_{22} & \overline{Q}_{26} \\ \overline{Q}_{61} & \overline{Q}_{62} & \overline{Q}_{66} \end{bmatrix} \tag{9}$$

For the case of on-axis laminae, where the fibers are parallel to the global x-axis direction, the equations for the stiffness matrix becomes:

$$[Q]_k = \begin{bmatrix} Q_{11} & Q_{12} & 0 \\ Q_{21} & Q_{22} & 0 \\ 0 & 0 & Q_{66} \end{bmatrix} \tag{10}$$

where

$$Q_{11} = \frac{E_1}{1 - \nu_{12}\nu_{21}} \tag{11}$$

$$Q_{12} = Q_{21} = \frac{\nu_{12}E_2}{1 - \nu_{12}\nu_{21}} \tag{12}$$

$$Q_{22} = \frac{E_2}{1 - \nu_{12}\nu_{21}} \tag{13}$$

$$Q_{66} = G_{12} \tag{14}$$

When the fibers of the lamina make an angle with the global x-axis direction, the lamina is called off-axis, and the stiffness matrix in Eq. (9) is populated based on the following equations

$$\overline{Q}_{11} = Q_{11}\cos^4\theta + 2(Q_{12} + 2Q_{66})\sin^2\theta\cos^2\theta + Q_{22}\sin^4\theta \tag{15}$$

$$\overline{Q}_{12} = \overline{Q}_{21} = (Q_{11} + Q_{22} - 4Q_{66})\sin^2\theta\cos^2\theta + Q_{22}\left(\sin^4\theta + \sin^4\theta\right) \tag{16}$$

$$\overline{Q}_{22} = Q_{11}\sin^4\theta + 2(Q_{12} + 2Q_{66})\sin^2\theta\cos^2\theta + Q_{22}\cos^4\theta \tag{17}$$

$$\overline{Q}_{16} = \overline{Q}_{61} = (Q_{11} - Q_{12} - 2Q_{66})\sin\theta\cos^3\theta + (Q_{12} - Q_{22} + 2Q_{66})\cos\theta\sin^3\theta \tag{18}$$

[1]It is often practice to account the global orientation of the laminate x,y,z, as coinciding with the local orientation 1,2,3 of the 0° fiber orientation. As a result, the 0° fibers, along local direction 1, are also along the x-direction of the laminate. Such laminae are called on-axis and any lamina of different orientation is termed off-axis.

$$\overline{Q}_{26} = \overline{Q}_{62} = (Q_{11} - Q_{12} - 2Q_{66})\cos\theta\sin^3\theta + (Q_{12} - Q_{22} + 2Q_{66})\sin\theta\cos^3\theta \quad (19)$$

$$\overline{Q}_{66} = (Q_{11} + Q_{22} - 2Q_{12} - 2Q_{66})\sin^2\theta\cos^2\theta + Q_{66}(\sin^4\theta + \cos^4\theta) \quad (20)$$

where $m = \cos\theta$ and $n = \sin\theta$.

To develop a constitutive equation for the kth lamina (Eq. (21)) stress distribution in each layer is related to the strain in each layer thought the stiffness matrix. The strain distribution depends on the mid-surface strains ($\varepsilon_{ij}{}^{o}$) and curvatures (κ_{ij}) developed in the laminate. Mid-surface strains and curvatures are laminate parameters, and therefore are the same for all laminae. They depend on the loading conditions, including the effects of thermal (α_{ij}) and hygral conditions (β_{ij}) in the laminate. Therefore, thermal strains and hygral effects, which are responsible for residual strains in the laminate, induced during manufacturing and curing, should be subtracted from the mid-surface strains and curvatures.

$$\begin{Bmatrix} \sigma_x \\ \sigma_y \\ \tau_{xy} \end{Bmatrix}_k = \left[\overline{Q}_k\right]\left(\begin{Bmatrix} \varepsilon_x^o \\ \varepsilon_y^o \\ \gamma_{xy}^o \end{Bmatrix} + z\begin{Bmatrix} \kappa_x \\ \kappa_y \\ \kappa_{xy} \end{Bmatrix} - \begin{Bmatrix} \alpha_x \\ \alpha_y \\ \alpha_{xy} \end{Bmatrix}_k \Delta T - \begin{Bmatrix} \beta_x \\ \beta_y \\ \beta_{xy} \end{Bmatrix}_k \overline{M}\right) \quad (21)$$

In order to build the stress and strain distributions in the laminate, there are three more important matrices in CLT that that need to be considered. These are: the Extension Stiffness Matrix, A_{ij}, the Extension-Bending Coupling Matrix, B_{ij}, and the Bending Stiffness Matrix, D_{ij}. These three matrices bring together the stiffness effects from each lamina, and consequently fiber orientation, always accounting for the position of each lamina in the laminate. The thickness, t, of each lamina is also an important factor in CLT and is accounted for through these three matrices. Each of these matrices determines the stress distribution of the laminate due to the different loading conditions that may be applied. Matrix A considers the tension-compression effects of longitudinal and transverse loading, matrix D considers the effects of bending moments, while matrix B couples the effects of both types of loading. Eqs. (22)–(24) refer to these three matrices, while Eq. (25), builds the relationship by calculating the normal forces and moments [6].

$$[A_{ij}] = \sum_{k=1}^{n} \left[\overline{Q}_{ij}\right]_k t_k \quad (22)$$

$$[B_{ij}] = \sum_{k=1}^{n} \left[\overline{Q}_{ij}\right]_k t_k \overline{z}_k \quad (23)$$

$$[D_{ij}] = \sum_{k=1}^{n} \left[\overline{Q}_{ij}\right]_k \left(t_k \overline{z}_k{}^2 + \frac{t_k{}^3}{12}\right) \quad (24)$$

$$\begin{Bmatrix} \widehat{N} \\ \cdots \\ \widehat{M} \end{Bmatrix} = \begin{pmatrix} A & \vdots & B \\ \cdots & \cdots & \cdots \\ B & \vdots & D \end{pmatrix}\begin{Bmatrix} \varepsilon^o \\ \cdots \\ \kappa \end{Bmatrix}^{2} \quad (25)$$

[2]All loading conditions, including thermal and hygral effects, are accounted for in $\widehat{N}$ and $\widehat{M}$.

3. How do composite materials fail?

The aforementioned concepts showed how the mechanical properties and the strength of the anisotropic material are evaluated. However, an important question to ask when designing FRP components against failure is how do composites fail.

Depending on how we evaluate the strength of a laminated composite; macroscopically or macroscopically, different modes of failure will become important to examine. While homogeneous materials, as is the case of metals, fail through crack initiation and propagation, often leading to fracture, composites accumulate damage and their strength degrades slowly. Degradation of the strength is often due to different failure modes and fatigue [2, 7, 8].

It has been briefly mentioned above that it is the reinforcing phase of FRPs that possesses the higher stiffness and strength. Therefore, it is the fibers that hold the load carrying capacity of the FRP composite. The load carrying capacity of composites is hidden in their fibers. As a result, the anisotropic nature of composites not only allows them to have different strengths in different directions, but also to have their constituents failing under different loads. In addition to this, due to the different fiber orientations and laminae positions in composite materials, these composites have a great advantage in failure. All lamina will not fail under the same loading conditions. Although one or more laminae may fail, the load may still be carried by the remaining strong laminae and the composite component may still be operational [9]. The matrix serves mostly as a mean of holding these fibers together and shaping the composite component, while its contribution to the load carrying capacity of the material is minimal. As a result, failure is usually aggravated when it involves degradation of the fiber load carrying capacity.

Looking at the constituent material level and their interface, i.e. microscopically, there are three failure modes that should be examined. These are:

1. Failure of the matrix phase, which is usually realized as in most homogeneous materials through crack initiation and propagation.

2. Failure at the reinforcing phase, which is the fracture of one or more fibers of the reinforcing phase.

3. Failure at the interface of the two constituents, referred to as debonding, where the fibers detach form the matrix material.

Each of the above failure modes, although they are responsible for the degradation of the composite mechanical properties, they affect the strength and performance of the material differently. Since fibers hold the most significant part of the composite's load carrying capacity, it is fracture of the fibers that can significantly impact and impair this capacity. It is the longitudinal axis of the fiber that holds their strength, and as a result, fracture of a fiber means a discontinuity in this strength along the longitudinal direction of the fiber, and consequently a degradation of this strength in the composite. Fractured fibers cannot be replaced, and therefore, this is a failure mode that will permanently degrade the strength of the material.

Although the majority of matrix materials in FRPs are thermosetting polymers, which means that a fractured matrix or debonding may not be repairable failure, the fact that the matrix holds an insignificant load carrying capacity, constitutes these two failure modes as lower intensity modes compared to fiber fracture.

At the laminate level, i.e. looking at the composite macroscopically, there exists another failure mode called delamination. Delamination is the separation, often also referred to as debonding, of consecutive plies. Delamination is one of the most common failure modes observed macroscopically and can significantly affect the performance of the laminated component, depending on the application the laminate is designed for. Delamination may be the result of poor manufacturing and it is for this reason that extra attention is being paid during the lamination of components to avoid contaminants entering between the layers, and also avoid the formation of voids due to air entrapment during the laminating process. Depending on the application, and polymer matrix of the FRP, delamination is a serious failure mode, but also one that may be repaired by a second curing process.

In both homogeneous and heterogeneous materials, there exist failure modes that are associated with the environment of the application the components are designed for. Different materials have different chemical properties and therefore are not always suited for all environments. Polymeric composites see many applications in highly corrosive environments [1], however, depending on the matrix or reinforcing phase they may not be suitable for applications at high temperatures, UV exposure, or high moisture content. As can be seen form the constitutive equation for FRP laminae, thermal stresses and moisture absorption will affect significantly the stresses developed in the lamina. Thermal gradients will not only affect the matrix material, but will also induce undesirable residual stresses in-between the lamina that can affect the strength of the composite, and lead to delamination or matrix fracture [6]. Similarly the moisture absorbing capacity of fibers is something that should be considered when selecting fiber-reinforcing phases. As an example, natural fibers, such as Hemp fibers, are high strength fibers investigated as a replacement to glass fibers in many applications in the automotive industry [10–12]. However, they are highly hygroscopic and their mechanical properties deteriorate faster due to moisture absorption [7, 12–16].

4. Predicting failure and damage in FRPs

The above failure modes may differ in the way that failure occurs, however, they all result in deteriorating the strength of the composite material. Examining the failure mechanism in each mode means working at the materials level, which is beyond the scope of this section. Delamination on the other hand, may be a failure mode that is often examined as a fracture mechanism, and for this reason will not be considered in the discussion that follows. However the failure result of the modes, to the extent that they affect the stress distribution in the composite, can be predicted by the following criteria.

Failure theories, or preferably failure criteria, such as the Tresca and von Mises are commonly used in ductile isotropic materials. These criteria, apart from the fact that are specific to ductile isotropic materials, they deal with parameters (stress and strain, respectively) in each direction separately [2, 6, 9]. The anisotropy of composite materials plays, as it has already been shown above, a significant role in the way that strength is built in the composite, and how failure can affect the composite's performance. Therefore, failure criteria should be selected based on their capacity to account for the interaction of stresses in different directions in the composite, where the material properties vary. The former type of failure criteria, as are the Tresca and von Mises, are referred to as non-interactive, because they account for each direction separately, while the latter, most suitable for composite materials, are referred to as interactive failure criteria. There exist various failure criteria that have been developed to address anisotropic materials, and specifically the case of FRPs. These criteria, although they account for the interaction of stresses, and consequently properties in different directions, cannot give an exact estimate or prediction for the stress conditions to cause failure. These criteria account for just the interaction of stresses irrespective of the failure mode or any other conditions (environmental, thermal, etc.), which may be responsible for composite failure. As a result, they are used as an estimate for the conditions at the onset of failure.

The majority of failure criteria are polynomial expansions, treating the stress tensor (σ_{ij}) as a means of characterizing the onset of failure in a composite material. As mentioned above, failure criteria can only give approximate estimates of the onset of failure, and have been developed based on comparisons to experimental data. The polynomial expansion is tailored to the case of transversely isotropic materials, which reduces significantly the number of material parameters required [2].

The criteria that will be discussed below compare the stress state in each lamina to failure stress under plane stress conditions and determine whether the lamina has failed or not. They are therefore, observing the composite material at the lamina level, and not down to the constituent material and interface level.

Tsai-Hill Failure Criterion:

$$\frac{\sigma_1^2}{X^2} - \frac{\sigma_1\sigma_2}{X^2} + \frac{\sigma_2^2}{Y^2} + \frac{\sigma_{12}^2}{S^2} \leq 1 \tag{26}$$

The Tsai-Hill criterion in Eq. (26) compares longitudinal (σ_1), transverse (σ_2), and shear stresses (σ_{12}) in each lamina to the ultimate longitudinal tensile and compressive (X and X′), transverse tensile and compressive (Y and Y′), and shear (S) ultimate strengths. A total of 5 parameters are required, while only 3 are involved in the equation. The criterion states that when the above is equal or greater than 1 failure has occurred.

Tsai-Wu Failure Criterion:

$$\begin{aligned}&\left(\frac{1}{X} - \frac{1}{X'}\right)\sigma_{11} + \left(\frac{1}{Y} - \frac{1}{Y'}\right)(\sigma_{22} + \sigma_{33}) + \frac{\sigma_{11}^2}{XX'} + \frac{1}{YY'}(\sigma_{22} + \sigma_{33})^2 + 2F_{12}\sigma_{11}(\sigma_{22} + \sigma_{33})\\&+\frac{1}{S'^2}\left(\sigma_{23}^2 - \sigma_{22}\sigma_{33}\right) + \frac{1}{S^2}\left(\sigma_{12}^2 + \sigma_{31}^2\right) \leq 1\end{aligned} \tag{27}$$

Similar to the Tsai-Hill, the Tsai-Wu criterion investigates failure at the lamina level stating that failure occurs when the above equation is equal to 1. As can be seen in Eq. (27), there are 6 constants involving the materials parameters of tensile and compressive strengths in the longitudinal and transverse directions, as well as shear strengths. Therefore, this criterion requires a total of 7 material parameters. The Tsai-Wu criterion terms can be evaluated by the assumption of uniaxial tension and compression results, which is based on experimental data [2, 6]. The interaction parameter (F_{12}) due to its interactive nature is often estimated from multiaxial stress data [2].

Both theories can give an estimated result as to when a lamina will fail. Tsai-Hill tends to overestimate failure, while Tsai-Wu tends to underestimate failure [17]. **Table 1** shows data using the above two failure criteria predicting the onset of failure in a 6-layer GFRP laminate under 3-point bending [17]. As can be observed form the data, the two theories for the case of the anti-symmetric laminate do not agree as to the exact magnitude of the criterion estimate, nor as to the possible first ply to fail. For this reason, more than one failure theories may be used in the design against failure of an FRP laminated component.

Lamina	Fiber Orientation	Tsai-Hill	Tsai-Wu
1	−45°	0.0539	−0.3632
2	0°	0.0989	−0.0027
3	90°	0.1289	−0.2924
4	90°	0.4688	−0.3083
5	90°	0.4688	−0.3083
6	90°	0.4688	−0.3083

Table 1. Example of failure prediction using Tsai-Hill and Tsai-Wu criteria in GFRP laminate.

As it was previously mentioned failure modes are often exhibited in different constituents of the composite, more specifically either the matrix or the fibers. The above example of how the Tsai-Hill and Tsai-Wu criteria over- and underestimate, respectively, the onset of failure, shows the approximate nature of the failure criteria, and the possible need to focus on the failing constituent. Of course, in the case where the fibers are significantly stiffer than the matrix, and failure is observed in the matrix, the fibers should be able to maintain the strength of the composite, while in the case of fiber fracture, the same will not hold true. Therefore, it is prudent that in situations of high anisotropy, an observation of failure in the matrix and fibers be made separately. The Christensen Criterion, also of the polynomial-expansion type, makes use of this differentiation between the two constituents, and requires 5 material parameters [2].

Christensen Failure Criterion:

Failure in the matrix

$$\left(\frac{1}{X}-\frac{1}{X'}\right)(\sigma_{22}-\sigma_{33})+\frac{1}{YY'}\left[(\sigma_{22}-\sigma_{33})^2+4{\sigma_{23}}^2\right]+\frac{\left({\sigma_{12}}^2+{\sigma_{31}}^2\right)}{S^2}\leq 1 \tag{28}$$

Failure in the fibers

$$\left(\frac{1}{X}-\frac{1}{X'}\right)(\sigma_{11})+\frac{\sigma_{11}{}^{2}}{XX'}-\frac{1}{4}\left(\frac{1}{X}+\frac{1}{X'}\right)^{2}\sigma_{11}(\sigma_{22}+\sigma_{33})\leq 1 \tag{29}$$

The above criteria, similar to all failure criteria developed for FRP composites, deal with the onset of failure in the lamina level. However, when designing against failure in a laminated FRP component, it is always useful to know under what conditions the component may fail and not which lamina will fail first. The advantage of composite materials is that they fail due to damage accumulation, which may be more delayed than catastrophic failure in an isotropic material, such as metal, due to cracking or yielding. The anisotropic nature of composites, with the fact that fibers hold the load carrying capacity of the composite material, allow the structure to continue performing under given conditions, although the properties in one or more laminae are being or have been degraded. Predicting when a composite structure will fail, meaning when all laminae have failed under one or more of the above criteria, is useful information and progressive damage is the process to follow. Progressive damage works by applying the above criteria of failure to determine which lamina will fail first, what we refer to as first-ply failure, and then continue to determine the next lamina, up to the last one. Every time a lamina fails, the properties of the composite are degraded and the failure of the remaining lamina is evaluated under new conditions of stress distribution and loading [2, 9].

Composite materials have been recognized as an optimum alternative to metals and have been replacing metals in many industries. For example, the automotive industry in an effort to decrease carriage weight, and consequently fuel emissions, has introduced composite suspensions. There are therefore, applications for composites that involve cyclic loading, which leads to fatigue failure. The above criteria may be helpful in estimating and approximating predictions of the stresses to initiate failure in a composite lamina, and together with progressive damage predict when the whole laminate will fail. However, the fatigue behavior of an FRP laminate should also be considered.

Cumulative Damage Theory calculates the damage caused during cyclic loading, as well as its accumulation during cyclic loading under various stress amplitudes [18]. There are two options when considering the concept of cumulative damage: a) calculating the residual strength of the component, being the instantaneous static strength maintained by the material after loading to stress levels that can cause damage, and b) estimating the damage accumulated in the material, using damage models [2].

As it has been shown above, for a composite to fail damage should be accumulated in each lamina and start degrading the lamina, and consequently laminate strength. This is also what happens during cyclic loading, where the strength of the material starts to decrease at a low rate in the early fatigue life, and at a faster rate close to the end, leading to possible failure [19].

Cumulative damage models do not focus on material data, rather on the loading conditions of the component in question. As a result, they predict damage accumulation in a general sense. Therefore, to decide upon one or more appropriate models to predict the fatigue life of a composite material, attention should be paid not to the material properties but to the type of stresses that cause failure in composites. Because of the stiff reinforcing phase holding the load carrying capacity of the lamina, it is higher stresses that will more likely degrade the strength of the material, than lower ones. As a result, cumulative damage models should focus on low cycle fatigue (LCF) where stresses are higher [20].

Three cumulative damage models are compared below for the case of GFRP laminates, reinforced with E-Glass fibers. This is a common FRP material considered in automotive applications as a substitute for steel. The three models require information on the number of cycles (n_i) under an applied stress (σ_i and σ_k), the number of cycles to failure (N_i) under this same applied stress, the ultimate strength of the material ($\sigma_{ultimate}$), the ratio of the applied stress to the ultimate material strength (S_k), and the number of repetitions of the loading cycle (C). Cumulative damage models denote failure, when the model equations equal to 1 [8, 18, 19, 21]. As a result, under the assumption that the material accumulates 100% damage to the full extent of its fatigue life, these models may be used to estimate the fatigue life, in cycles to failure (N_i), by setting the models to equal 1.

Palmgren-Miner:

$$\left(\sum_{i=1}^{n}\frac{n_i}{N_i}\right)C = 1 \tag{30}$$

The Palmgren-Miner cumulative damage model is maybe one of the simplest and most commonly used models in metal fatigue. The model defines the damage accumulated in the material in the form of life fractions. Each fraction is a percentage of composite life consumed during the cyclic loading application [18, 22]. When the sum of all fractions equals 1, there is no remaining residual life to be expended, and the material is assumed to have failed. The Palmgren-Miner model does not account for the loading sequence in the case of different applied stress amplitudes.

Broutman-Sahu:

$$\left(\sum_{i=1}^{n}\frac{(\sigma_{ultimate}-\sigma_i)}{(\sigma_{ultimate}-\sigma_{i+1})}\frac{n_i}{N_i}\right)C = 1 \tag{31}$$

Hashin-Rotem:

$$\left(\sum_{k=1}^{i-1}\left(\frac{n(k-1)}{N(k-1)}\right)^{\frac{1-S_k}{1-S_{k-1}}}+\frac{n_i}{N_i}\right)C = 1 \tag{32}$$

$$S_k = \frac{\sigma_k}{\sigma_{Ultimate}} \tag{32a}$$

$$S_{k-1} = \frac{\sigma_{k-1}}{\sigma_{Ultimate}} \tag{32b}$$

Although the Palmgren-Miner and Hashin-Rotem have been initially designed as cumulative damage models for metals, they have been both used for FRP, and more specifically GFRP fatigue life predictions. Hashin-Rotem is designed as a two-stress level loading damage model, and can be expanded to multi-stress level loading using damage curve families that represent residual lifetimes and considering that equivalent residual lives are expended by components undergoing different loading schemes [18]. The Broutman-Sahu model was developed and tested on GFRP laminates.

The linearity and non-linearity of the above models is important and is determined based on the required parameters for their calculation [8]. As a result, Palmgren-Miner is a linear stress-independent model, Broutman-Sahu a linear stress-dependent model, and Hashin-Rotem a non-linear stress-dependent model.

Similar to the failure criteria, these models give an approximation of the accumulated damage during cyclic loading, and consequently when used to estimate cycles to failure, the fatigue life of the laminate. To evaluate the applicability of these models to composite materials, and choose those that predict the fatigue life of a GFRP laminate more accurately, calculations of the above models were compared to fatigue life experiments [23–24]. In order to estimate the probability to failure for the GFRP laminate using each model a standard two-parameter Weibull analysis was followed.

Figure 3, shows comparison of the tree models for an E-glass GFRP beam cycled between 256 and 560 MPa. All three models give similar results for the cumulative distribution of damage in the GFRP laminate. Specifically, the damage predicted by the Palmgren-Miner and Broutman-Sahu models is almost identical, and the two linear models show a constant probability of failure of 19%, at low mean stresses up to 280 MPa. All three models predict complete failure at a mean stress of 560 MPa. This mean stress level corresponds to a maximum stress of 1.1 GPa, which exceedsthe ultimate strength of the material (**Figure 4**) [23–24].

Comparing the results of the three models to experimental fatigue data, it can be concluded that similar to the failure criteria, cumulative damage models can be used to approximately

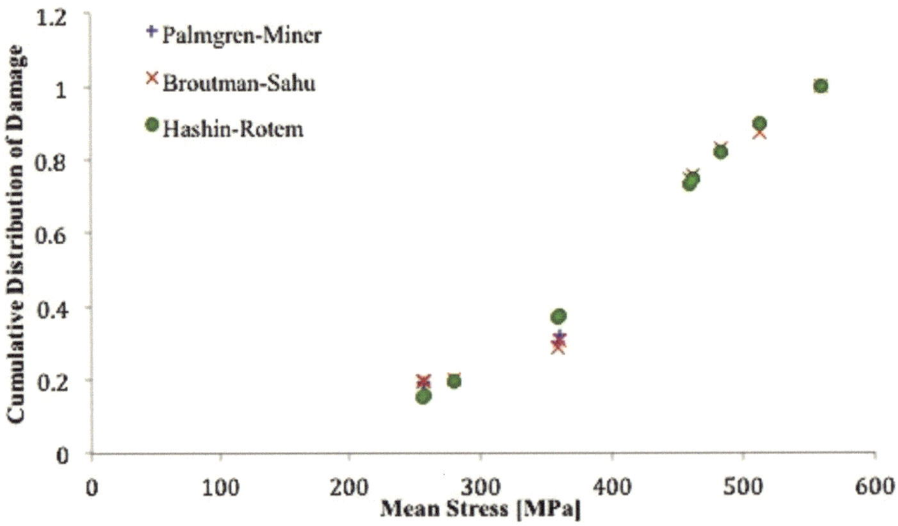

Figure 3. Cumulative distribution of damage vs. mean stress in E-glass fiber/epoxy laminate.

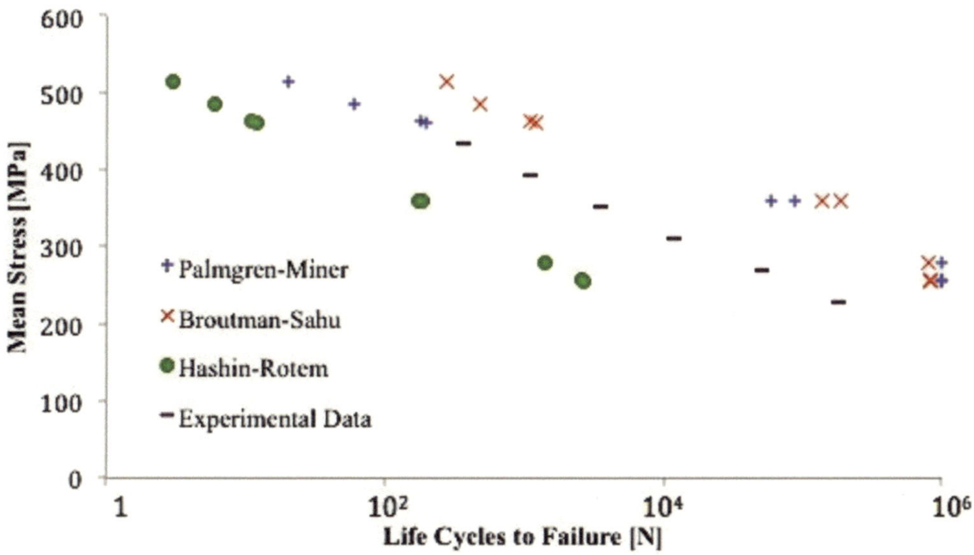

Figure 4. Mean stress vs. life to failure in E glass fiber/epoxy.

predict the fatigue life of a composite component, without getting accurate life predictions. In the case of the GFRP of **Figure 4**, the two linear models give overestimated predictions of fatigue life, and the non-linear model a significantly underestimated prediction, when model data is compared to experimental data. The linear stress-dependent model, is the one to estimate a predicted life having closer agreement with experimental results.

5. Conclusion

The versatility of composite materials, and specifically the case of FRPs, have made laminates ideal alternative structures to metallic components. The ability to design against failure by optimizing the stacking sequence of laminates while at the same time designing not just a strong component but also one with desirable stiffness, have opened new horizons to the use of FRPs.

This chapter discussed the theory behind building constitutive relationships for laminae and consequently laminate structures. Using the stress information from the CLT, it was shown how to estimate the onset of failure under different loadings. Of course, when designing against failure it is prudent to consider and consult more than one failure criteria, as they do not give exact predictions for the onset of failure, rather approximations. By considering one or more models, bounds can be drawn to limit the onset of failure conditions. The same holds true for the case of fatigue life, where cumulative damage models also tend to either overestimate or underestimate the fatigue life of laminates.

The above criteria and models, however, besides their predictive character, which to a large degree is due to the anisotropic nature of the laminates, as well as to the fact that they consider only mechanical failure in the composite, can be applied to laminates of more than one constituents, as well as hybrid composite laminates. Therefore, if the optimization of a component requires more than one matrix and/or reinforcing phase, the above analysis could be followed to get estimates for the onset of failure and fatigue life of the component. The same holds true for hybrid laminates, where a third material, usually a reinforcing phase, other than fibers, for example a metallic reinforcement, is introduced. As a result, the above criteria and models can see applications in current designs of new composite materials, as the components built in n-d printing where the structures have not only varying material orientations, but also material properties.

The failure criteria and cumulative damage models, become therefore, an invaluable set of tools to the design against failure of laminate components. This design with the use of CLT can provide optimum laminates that have the desired stacking sequence to optimize mechanical properties and weight, as well as cost requirements. However, the discussion of this chapter did not account for failure that is not mechanical. The failure modes discussed concentrate on matrix or fiber cracking, debonding, and delamination, which was not examined at the laminate level. However, thermal and hygral effects are only accounted for in CLT and are not part of the failure criteria presented in this chapter. As a result, when failure is due to chemical degradation of the matrix under UV rays, or creep which should involve the time parameter, failure should be considered using different models which were not the scope of this chapter.

Nomenclature

E_{ijkl}	Young's Modulus
G_{ij}	Shear Modulus
K	Bulk Modulus
f	Volume Fraction
v_{ij}	Poisson's Ratio
σ_{ij}	Stress Tensor
ε_{kl}	Strain Tensor
κ_{ij}	Curvature
$\varepsilon_{ij}{}^{o}$	Mid-Surface Strains
α_{ij}	Coefficient of Thermal Expansion
β_{ij}	Hygroscopic Coefficient
$\overline{Q}$	Stiffness Matrix
A_{ij}	Extension Stiffness Matrix

B_{ij}	Extension-Bending Coupling Matrix
D_{ij}	Bending Stiffness Matrix
z	Position of layer in laminate
X and X′	Longitudinal Tensile and Compressive Strength.
Y and Y′	Transverse Tensile and Compressive Strength.
S	Shear Strength
n_i	Number of cycles under applied stress
N_i	Cycles to failure
C	Repetitions of cyclic loading

Author details

Roselita Fragoudakis

Address all correspondence to: fragoudakisr@merrimack.edu

Merrimack College, North Andover, MA, USA

References

[1] Gürdal Z, Haftka RT, Hajela P. Design and Optimization of Laminated Composite Materials. 1st ed. Wiley-Interscience; 1999

[2] Christense RM. Mechanics of Composite Materials. Mineola, NY, USA: Dover; 2005

[3] Barbero EJ. Introduction to Composite Materials Design, CRC Press. 2nd ed. FL, USA: Boca Raton; 2010

[4] Dvorak G. Micromechanics of Composite Materials. Troy, NY, USA: Springer; 2013

[5] Vasiliev VV, Morozov EV. Advanced Mechanics of Composite Materials and Structural Elements. 3rd ed. Waltham, MA, USA: Elsevier; 2013

[6] Staab GH. Laminar Composites. Butterworth-Heinemann, Kingston, OX, UK; 1999

[7] Agarwal BD, et al. Analysis and Performance of Fiber Composites. West Sussex, UK: John Wiley & Sons; 2006

[8] Owen MJ. Fatigue damage in glass-fiber-reinforced plastics. In: Broutman LJ, editor. Composite Materials. Academic Press; 1974. pp. 313-340

[9] Jones RM. Mechanics of Composite Materials. 2nd ed. Philadelphia, PA, USA: Taylor & Francis, Inc.; 1999

[10] Wambua P, et al. Natural fibers: Can they replace glass in glass fiber reinforced plastics? Composites Science and Technology. 2003;**63**:1259-1264

[11] Corniere-Nicollier T, et al. Life cycle assessment of Biofibres replacing glass fibers as reinforcement in plastics. Resources, Conservation and Recycling. 2001;**33**:267-287

[12] Joshi SV, et al. Are natural fiber composites environmentally superior to glass fiber reinforced composites? Composites: Part A. 2005;**35**:371-376

[13] Dhakal HN, et al. Effect of water absorption on the mechanical properties of hemp fibre reinforced unsaturated polyester composites. Composites Science and Technology. 2007; **67**:1674-1683

[14] Wotzel K, et al. Life cycle studies on hemp fiber reinforced components and ABS for automotive parts. Angewandte Makromolekulare Chemie. 1999;**272**:121-127

[15] Wang W. Study of moisture absorption in natural fiber plastic composites. Composites Science and Technology. 2006;**66**:379-386

[16] Fragoudakis R, et al. A computational analysis of the energy harvested by GFRP and NFRP laminated beams under cyclic loading. Procedia Engineering. 2017;**200**:221-228

[17] Fragoudakis R. Predicting the optimum Stackign sequence of fiber reinforced plastic laminated beams under bending. In: SAMPE Seattle 2017; 22 May - 24-05-2017; Seattle, WA. 2017

[18] Hashin Z, Rotem A. A cumulative damage theory of fatigue failure. Journal of Materials Science and Engineering. 1978;**34**:147-160

[19] Broutman LJ, Sahu SA. Progressive damage of glass reinforced plastic during fatigue. In: 22nd Annual technical Conference, Reinforced Plastics/Composite Div.; 1969

[20] Salkind MJ. Fatigue of composites, composite materials: Testing and design. In: ASTM STP 497: American Society for Testing Materials; 1972. p. 143–169

[21] Epaarachi JA. A study on estimation of damage accumulation of glass fibre reinforced plastic (GFRP) composited under block loading situation. Composite Structures. 2006;**75**: 88-92

[22] Suresh S. Fatigue of Materials. Cambridge, UK: Cambridge University Press; 1998

[23] Fragoudakis R, Saigal A. Predicting the fatigue life in steel and glass fibre reinforced plastics using damage models. Journal of Materials Science and Applications. 2011;**2**: 596-604

[24] Fragoudakis R, Saigal A. Using damage models to predict fatigue in steel and glass fibre reinforced plastics. Journal of Materials Science and Engineering with Advanced technologies. 2011;**3**:53-65

Chapter 7

Failure Rate Analysis

Fatemeh Afsharnia

Additional information is available at the end of the chapter

http://dx.doi.org/10.5772/intechopen.71849

Abstract

Failure prediction is one of the key challenges that have to be mastered for a new arena of fault tolerance techniques: the proactive handling of faults. As a definition, prediction is a statement about what will happen or might happen in the future. A failure is defined as "an event that occurs when the delivered service deviates from correct service." The main point here is that a failure refers to misbehavior that can be observed by the user, which can either be a human or another computer system. Things may go wrong inside the system, but as long as it does not result in incorrect output (including the case that there is no output at all) there is no failure. Failure prediction is about assessing the risk of failure for some time in the future. In my approach, failures are predicted by analysis of error events that have occurred in the system. As, of course, not all events that have occurred ever since can be processed, only events of a time interval called embedding time are used. Failure probabilities are computed not only for one point of time in the future, but for a time interval called prediction interval.

Keywords: failure, component, analysis, reliability, probability

1. Introduction

Failure prediction is one of the key challenges that have to be mastered for a new arena of fault tolerance techniques: the proactive handling of faults. As a definition, prediction is a statement about what will happen or might happen in the future. A failure means "an occurrence that happens when the delivered service gets out from correct service."

The main point here is that a failure derives of misbehavior that can be observed by the operator, which can either be a human or another computer system. Some things may go wrong inside the system, but as long as it does not eventuate in incorrect output (such as the system that there is no output at all) the system can run without failure. Failure prediction is about evaluation the risk of failure for some times in the future. In my viewpoint, analysis of

INTECH
open science | open minds

© 2017 The Author(s). Licensee InTech. This chapter is distributed under the terms of the Creative Commons Attribution License (http://creativecommons.org/licenses/by/3.0), which permits unrestricted use, distribution, and reproduction in any medium, provided the original work is properly cited. (cc) BY

error events that have occurred in the system can be called failure prediction. To compute breakdown probabilities, not only one point of time in the future, but a time interval called prediction interval are considered, simultaneously.

Failure rates and their projective manifestations are important factors in insurance, business, and regulation practices as well as fundamental to design of safe systems throughout a national or international economy. From an economic view point, inaction owing to machinery failures as a consequence of downtimes can be so costly. Repairs of broken down machines are also expensive, because the breakdowns consume resources: manpower, spare parts, and even loss of production. As a result, the repair costs can be considered as an important component of the total machine ownership costs. Traditional maintenance policies include corrective maintenance (CM) and preventive maintenance (PM). With CM policy, maintenance is performed after a breakdown or the occurrence of an obvious fault. With PM policy, maintenance is performed to prevent equipment breakdown. As an example, it is appeared that in developing countries, almost 53% of total machine expenses have spent to repair machine breakdowns whereas it was 8% in developed countries, that founding the effective and practicable repair and maintenance program could decreased these costs up to 50%.

The complex of maintenance activities, methodologies and tools aim to obtain the continuity of the productive process; traditionally, this objective was achieved by reviewing and substituting the critical systems or through operational and functional excess in order to guarantee an excess of productive capacity. All these approaches have partially emerged inefficiencies: redundant systems and surplus capacity immobilize capitals that could be used more Affordable for the production activities, while accomplishing revision policies very careful means to support a rather expensive method to achieve the demand standards. The complex of maintenance activities is turned from a simple reparation activity to a complex managerial task which main aim is the prevention of failure. An optimal maintenance approach is a key support to industrial production in the contemporary process industry and many tools have been developed for improving and optimizing this task.

The majority of industrial systems have a high level of complexity, nevertheless, in many cases, they can be repaired. Moreover historical and or benchmarking data, related to systems failure and repair patterns, are difficult to obtain and often they are not enough reliable due to various practical constraints. In such circumstances, it is evident that a good RAM analysis can play a key role in the design phase and in any modification required for achieving the optimized performance of such systems. The assessing of components reliability is a basic sight for appropriate maintenance performance; available reliability assessing procedures are based on the accessibility of knowledge about component states. Nevertheless, the states of component are often uncertain or unknown, particularly during the early stages of the new systems development. So for these cases, comprehending of how uncertainties will affect system reliability evaluation is essential. Systems reliability often relies on their age, intrinsic factors (dimensioning, components quality, material, etc.) and use conditions (environment, load rate, stress, etc.). The parameter defining a machine's reliability is the failure rate (λ), and this value is the characteristic of breakdown occurrence frequency. In this context, failure rate analysis constitute a strategic method for integrating reliability, availability and maintainability, by using methods, tools and engineering techniques (such as Mean Time to Failure, Equipment down Time and System Availability values) to identify and quantify equipment and system failures that prevent the achievement of its objectives. At first we define common words related to failure rate:

- Failure

A failure occurs when a component is not available. The cause of components failure is different; they may fail due to have been randomly chosen and marked as fail to assess their effect, or they may fail because any other component that were depending on else has brake down. In reliability engineering, a Failure is considered to event when a component/system is not doing its favorable performance and considered as being unavailable.

- Error

In reliability engineering, an error is said a misdeed which is the root cause of a failure.

- Fault

In reliability engineering, a fault is defined as a malfunction which is the root cause of an error. But within this chapter, we may refer to a component failure as a fault that may be conducted to the system failure. This is done where there is a risk of obscurity between a failure which is occurring in intermediate levels (referred to as a Fault) and one which is occurring finally (referred to as Failure).

2. Failure Rate

The reliability of a machine is its probability to perform its function within a defined period with certain restrictions under certain conditions. The reliability is the proportional expression of a machine's operational availability; therefore, it can be defined as the period when a machine can operate without any breakdowns. The equipment reliability depends to failures frequency, which is expressed by MTBF[1]. Reliability predictions are based on failure rates. Failure intensity or $\lambda(t)$[2] can be defined as "the foretasted number of times an item will break down in a determined time period, given that it was as good as new at time zero and is functioning at time t". This computed value provides a measurement of reliability for an equipment. This value is currently described as failures per million hours (f/mh). As an example, a component with a failure rate of 10 fpmh would be anticipated to fail 10 times for 1 million hours time period. The calculations of failure rate are based on complex models which include factors using specific component data such as stress, environment and temperature. In the prediction model,

[1]Mean time between failures.

[2]Conditional failure rate.

assembled components are organized serially. Thus, failure rates for assemblies are calculated by sum of the individual failure rates for components within the assembly. The MTBF was determined using Eq. (1). Failure rate which is equal to the reciprocal of the mean time between failures (MTBF) defined in hours (λ) was calculated by using Eq. (2) [1].

$$MTBF = \frac{T}{n} \tag{1}$$

$$\lambda = \frac{1}{MTBF} \tag{2}$$

where, MTBF is mean time between failures, h; T is total time, h; n is number of failures; λ is failure rate, failures per 10^n h.

There are some common basic categories of failure rates:

- Mean Time Between Failures (MTBF)
- Mean Time To Failure (MTTF)
- Mean Time To Repair (MTTR)
- Mean Down Time (MDT)
- Probability of Failure on Demand (PFD)
- Safety Integrity Level (SIL)

2.1. Mean time between failures (MTBF)

The basic measure of reliability is mean time between failures (MTBF) for repairable equipment. MTBF can be expressed as the time passed before a component, assembly, or system break downs, under the condition of a constant failure rate. On the other hand, MTBF of repairable systems is the predicted value of time between two successive failures. It is a commonly used variable in reliability and maintainability analyses. MTBF can be calculated as the inverse of the failure rate, λ, for constant failure rate systems. For example, for a component with a failure rate of 2 failures per million hours, the MTBF would be the inverse of that failure rate, λ, or:

$$MTBF = \frac{1}{\lambda}$$

or

$$\frac{1}{2\ failures/10^6\ hours} = 500{,}000\ hours/failure$$

> NOTE: Although MTBF was designed for use with repairable items, it is commonly used for both repairable and non-repairable items. For non-repairable items, MTBF is the time until the first (an only) failure after t_0.

http://dx.doi.org/10.5772/intechopen.71849

$$\text{Failure rate } (\lambda) = \frac{k}{T}$$

3. Units

Any unit of time can be mentioned as failure rate unit, but hours is the most common unit in practice. Other units included miles, revolutions, etc., which can also replace the time units.

In engineering notation, failure rates are often very low because failure rates are often expressed as failures per million (10^{-6}), particularly for individual components.

The failures in time (FIT) rate for a component is the number of failures that can be occurred in one billion (10^9) use hours. (e.g., 1000 components for 1 million hours, or 1 million components for each 1000 hours, or some other combination). Semiconductor industry currently used this unit.

Example 1 If we aim to estimate the failure rate of a certain component, we can carry out this test. Suppose each one of 10 same components are tested until they either break down or reach 1000 hours, after this time the test is completed for each component. The results are shown in **Table 1** as follows:

Component	Hours	Failure
1	950	Failed
2	720	No failure
3	467	No failure
4	635	No failure
5	1000	Failed
6	602	No failure
7	1000	Failed
8	582	No failure
9	940	Failed
10	558	Failed
Totals	7454	5

$Estimated\ failure\ rate = \frac{5\ failures}{7454\ hours} = 0.00067\ \frac{failures}{hour} = 670 \times 10^{-6}\ \frac{failures}{hour}$, or 670 failures/million use hours.

Table 1. Components failures during use hours.

Example 2 If a tractor be operated 24 hours a day, 7 days a week, so it will run 6540 hours for 1 year and at which time the MTBF number of a tractor be 1,050,000 hours:

$\frac{1{,}050{,}000\ hours}{6540\ hours/year} = 160.55\ years$; then the reciprocal of 160.55 years should be taken.

$$\frac{1\ failure}{160.55\ years} \times 100\% = 0.62\%$$

In the average year, we can expect to fail about 0.62% of these tractors.

Example 3 Now assuming a tractor be operated at 6320 hours a year and at which time the MTBF number of this be 63,000 hours.

$\frac{63{,}000\ hours}{6320\ hours/year} = 9.968\ years$ then the reciprocal of 9.968 years should be taken.

$\frac{1\ failure}{9.968\ years} \times 100\% = 10.032\%$ in the average year, we can expect to fail about 10.032% of these tractors.

You assume, we let the identical tractor run 24 hours a day, 7 days a week:

$$\frac{63{,}000\ hours}{8760\ hours/year} = 7.1917\ years$$

$\frac{1\ failure}{7.1917\ years} \times 100\% = 13.9\%$, i.e., ~13.9% of these tractors may break down in the average year.

3.1. Mean time to failure (MTTF)

One of basic measures of reliability is mean time to failure (MTTF) for non-repairable systems. This statistical value is defined as the average time expected until the first failure of a component of equipment. MTTF is intended to be the mean over a long period of time and with a large number of units. For constant failure rate systems, MTTF can calculated by the failure rate inverse, 1/λ. Assuming failure rate, λ, be in terms of failures/million hours, MTTF = 1,000,000/failure rate, λ, for components with exponential distributions. Or:

$$MTTF = \frac{1}{\lambda\ failures/10^6 hours} \tag{3}$$

For repairable systems, MTTF is the anticipated time period from repair to the first or next break down.

3.2. Mean time to repair (MTTR)

Mean time to repair (MTTR) can described as the total time that spent to perform all corrective or preventative maintenance repairs divided by the total of repair numbers. It is the anticipated time period from a failure (or shut down) to the repair or maintenance fulfillment. This is a term that typically only used in repairable systems.

Four failure frequencies are commonly used in reliability analyses:

Failure Density f(t)- The failure density of a component or system means that first failure what is likely to occur in the component or system at time t. In such cases, the component or system was running at time zero.

Failure Rate or r(t)- The failure rate of a component or system is expressed as the probability per unit time that the component or system experiences a failure at time t. In such cases, the component or system was using at time zero and has run to time t.

Conditional failure rate or conditional failure intensity λ(t)– The conditional failure rate of a component or system is the probability per unit time that a failure occurs in the component or system at time t, so the component or system was operating, or was repaired to be as good as new, at time zero and is operating at time t.

Unconditional failure intensity or failure frequency ω(t)– The definition of the unconditional failure intensity of a component or system is the probability per unit time when the component or system fail at time t. In such cases, the component or system was using at time zero. The following relations (4) exist between failure parameters [2].

$$R(t) + F(t) = 1 \tag{4}$$
$$f(t) = \frac{dF(t)}{dt}$$
$$F(t) = \int_0^t f(u).du$$
$$r(t) = \frac{f(t)}{1 - F(t)}$$
$$R(t) = e^{-\int_0^t r(u).du}$$
$$F(t) = 1 - e^{-\int_0^t r(u).du}$$
$$f(t) = r(t).e^{-\int_0^t r(u).du}$$

The difference between definitions for failure rate r(t) and conditional failure intensity λ(t) refers to first failure that the failure rate specifies this for the component or system rather than any failure of the component or system. Especially, if the failure rate being constant at considered time or if the component is non-repairable. These two quantities are same. So:

$$\lambda(t) = r(t) \quad for\ non-repairable\ components$$
$$\lambda(t) = r(t) \quad for\ constant\ failure\ rates$$
$$\lambda(t) \neq r(t) \quad for\ the\ general\ case$$

The conditional failure intensity (CFI) λ(t) and unconditional failure intensity ω(t) are different because the CFI has an additional condition that the component or system has survived to time t. The equation (5) mathematically showed the relationship between these two quantities.

$$\omega(t) = \lambda(t)[1 - Q(t)] \tag{5}$$

3.3. Constant failure rates

If the failure rate is constant then the following expressions (6) apply:

$$\begin{aligned} R(t) &= e^{-\lambda t} \\ F(t) &= 1 - e^{-\lambda t} \\ f(t) &= \lambda.e^{-\lambda t} \end{aligned} \tag{6}$$

As can be seen from the equation above, a constant failure rate results in an exponential failure density distribution.

3.4. Mean down time (MDT)

In organizational management, mean down time (MDT) is defined as the mean time that a system is not usable. This includes all time such as repair, corrective and preventive maintenance, self-imposed downtime, and any logistics or administrative delays. The MDT and MTTR (mean time to repair) are difference due to the MDT includes any and all delays involved; MTTR looks particularly at repair time.

Sometimes, Mean Time To Repair (MTTR) is used in this formula instead of MDT. But MTTR may not be the identical as MDT because:

- Sometimes, the breakdown may not be considered after it has happened
- The decision may be not to repair the equipment immediately
- The equipment may not be put back in service immediately it is repaired

If you used MDT or MTTR, it is important that it reflects the total time for which the equipment is unavailable for service, on the other hands the computed availability will be incorrect.

In the process industries, MTTR is often taken to be 8 hours, the length of a common work shift but the repair time really might be different particularly in an installation.

3.5. Probability of failure on demand (PFD)

PFD is probability of failure on demand. The design of safety systems are often such that to work in the background, monitoring a process, but not doing anything until a safety limit is overpassed when they must take some action to keep the process safe. These safety systems are often known as emergency shutdown (ESD) systems.

PFD means the unavailability of a safety task. If a demand to act occurs after a time, what is the probability that the safety function has already failed? As you might expect, the PFD equation looks like the equation (7) for general unavailability [3]:

$$\text{PFDavg} \approx \lambda_{DU}\ \text{MDT} \tag{7}$$

Note that we talk about PFDavg here, the mean probability of failure on demand, which is really the correct term to use, since the probability does change over time—the failure probability of a system will relied on how long ago you tested it.

λ_{DU} is the failure rate of dangerous undetected failures. We are not counting any failures that are guessed to be "safe," perhaps because they cause the process to shut down, only those failures which remain hidden but will fail the operation of the safety function when it is called upon.

This is essential as it assures us not to suppose that a safety-related product is generally more reliable than a general purpose product. The aim of safety-related product design is to have especially low failure rate of the safety task, but its total failure rate (MTBF) may not be so efficient.

So, the MDT for a safety function is defined as a dangerous undetected failure will not be obvious until either a demand comes along or a proof test would be revealed it.

Suppose we proof test our safety function every year or two, say every T_1 hours. The safety function is equally likely to fail at any time between one proof test and the next, so, on average it is down for $T_1/2$ hours.

From this we get the simplest form of PFD calculation for safety functions [3]:

$$PFDavg \approx ½ \lambda_{DU} T_1 \tag{8}$$

3.6. SIL

Under reliability engineering, SIL is one of the most abused terms. "SIL" is often used to mention that an equipment or system show better quality, higher reliability, or some other desirable feature. It does not. SIL actually means safety integrity level and has a range between 1 and 4. It is applied to depict the safety protection degree required by a process and finally the safety reliability of the safety system is essential to obtain that protection. SIL4 shows the highest level of safety protection and SIL1 is the lowest.

Many products are demonstrated by "SIL" rated. This means that they are appropriate for use in safety systems. In fact, if this is true, it relies on a lot of detail, which is beyond the scope of this chapter. But remember that even when a product indeed matches with "SIL" needs that are only reminding you that it will do a definite job in a safety system. This safety reliability may be high, but its general reliability may not be, as mentioned in the prior section.

Useful to remember

- If an item works for a long time without breakdown, it can be said is highly reliable.
- If an item does not fail very often and, when it does, it can be quickly returned to service, it would be highly available.
- If a system is reliable in performing its safety function, it is considered to be safe. The system may fail much more frequently in modes that are not considered to be dangerous.
- Finally, a safety system may be has lower MTBF in total than a non-safety system performing a similar function.
- "SIL" does not mean a guarantee of quality or reliability, except in a defined safety context.
- MTBF is a measure of reliability, but it is not the expected life, the useful life, or the average life.
- Calculations of reliability and failure rate of redundant systems are complex and often counter-intuitive.

4. Failure types

Failures generally be grouped into three basic types, though there may be more than one cause for a particular case. The three types included: early failures, random failures and wear-out failures. In the early life stage, failures as infant mortality often due to defects that escape the manufacturing process. In general, when the defective parts fail leaving a group of defect free products, the number of failures caused by manufacture problems decrease. Consequently the early stage failure rate decreases with age. During the useful life, failures may related to freak accidents and mishandling that subject the product to unexpected stress conditions. Suppose the failure rate over the useful life is generally very low and constant. As the equipment reaches to the wear-out stage, the degradation of equipment is related to repetitious or constant stress conditions. The failure rate during the wear-out stage increases dramatically as more and more occurs failure in equipment that caused by wear-out failures. When plotting the failure rate over time as illustrated in **Figure 1**, these stages make the so-called "bath tub" curve.

4.1. Early life period

To ensure the integrity of design, we used many methods. Some of the design techniques include: burn-in (to stress devices under constant operating conditions); power cycling (to stress devices under the surges of turn-on and turn-off); temperature cycling (to mechanically and electrically stress devices over the temperature extremes); vibration; testing at the thermal destruct limits; highly accelerated stress and life testing; etc. Despite usage of all these design tools and manufacturing tools such as six sigma and quality improvement techniques, there will still be some early failures because we will not able to control processes at the molecular level. There is always the risk that, although the most up to date techniques are used in design and manufacture, early breakdowns will happen. In order to remove these risks — especially in newer product consumes some of the early useful life of a module via stress screening. The start of operating life in initial peak represents the highest risk of failure; since in this

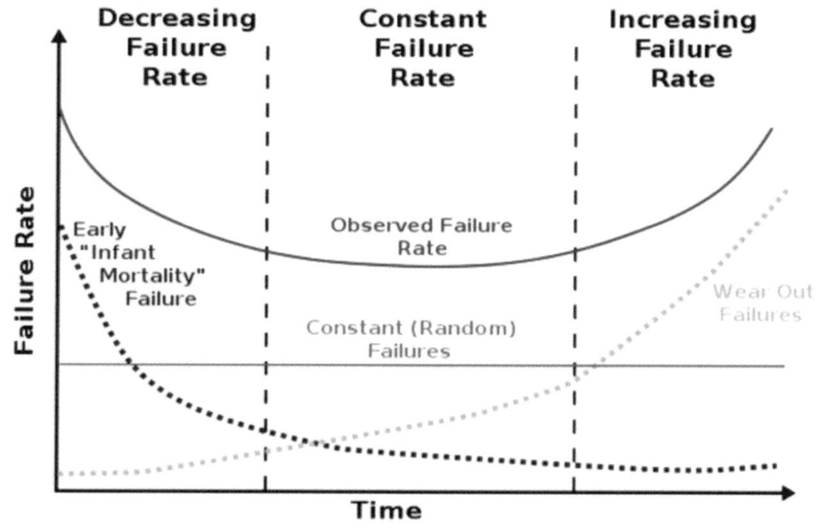

Figure 1. Bathtub curve for an ideal machine or component.

technique, the units are allowed to begin their somewhere closer to the flat portion of the bathtub curve. Two factors included burn in and temperature cycling consumed the operating life. The amount of screening needed for acceptable quality is a function of the process grade as well as history. M-Grade modules are screened more than I-Grade modules, and I-Grade modules are screened more than C-Grade units.

4.2. Useful life period

The maturity of product is caused that the weaker units extinct, the failure rate nearly shows a constant trend, and modules have entered what is considered the normal life period. This period is characterized by a relatively constant failure rate. The length of this period is related to the product or component system life. During this period of time, the lowest failure rate happens. Notice how the amplitude on the bathtub curve is at its lowest during this time. The useful life period is the most common time frame for making reliability predictions.

4.3. MTBF vs. useful life

Sometimes MTBF is Mistakenly used instead of component's useful life. Consider, the useful life of a battery is 10 hours and the measure of MTBF is 100,000 hours. This means that in a set of 100,000 batteries, there will be about one battery failure every 1 hour during their useful lives.

Sometimes these numbers are so much high, it is related to the basis calculations of failure rate in usefulness period of component, and we suppose that the component will remain in this stage for a long period of time. In the above example, wear-out period decreases the component life, and the usefulness period becomes much smaller than its MTBF so there is not necessarily direct correlation between these two.

Consider another example, there are 15,000 18-year-old humans in the sample. Our investigation is related to 1 year. During this period, the death rate became 15/15,000 = 0.1%/year. The inverse of the failure rate or MTBF is 1/0.001 = 1000. This example represents that high MTBF values is different from the life expectancy. As people become older, more deaths occur, so the best way to calculate MTBF would be monitor the sample to reach their end of life. Then, the average of these life spans are computed. Then we approach to the order of 75–80 which would be very realistic.

4.4. Wear-out period

As fatigue or wear-out occurs in components, failure rates increasing high. Power wear-out supplies is usually due to the electrical components breakdown that are subject to physical wear and electrical and thermal stress. Furthermore, the MTBFs or FIT rates calculated in the useful life period no longer apply in this area of the graph. A product with a MTBF of 10 years can still exhibit wear-out in 2 years. The wear-out time of components cannot predict by parts count method. Electronics in general, and Vicor power supplies in particular, are designed so that the useful life extends past the design life. This way wear-out should never occur during the useful life of a module.

4.5. Failure sources

There are two major categories for system outages: 1. Unplanned outages (failure) and 2. Planned outages (maintenance) that both conducted to downtime. In terms of cost, unplanned and planned outages are compared but use the redundant components maybe mitigate it. The planned outage usually has a sustainable impact on the system availability, if their schematization be appropriate. They are mostly happen due to maintenance. Some causes included periodic backup, changes in configuration, software upgrades and patches can caused by planned downtime. According to prior research studies 44% of downtime in service providers is unscheduled. This downtime period can spent lots of money.

Another categorization can be:

- Internal outage
- External outage

Specification and design flaws, manufacturing defects and wear-out categorized as internal factors. The radiation, electromagnetic interference, operator error and natural disasters can considered as external factors. However, a well-designed system or the components are highly reliable, the failures are unavoidable, but their impact mitigation on the system is possible.

4.6. Failure rate data

The most common ways that failure rate data can be obtained as following:

- Historical data about the device or system under consideration.

  Many organizations register the failure information of the equipment or systems that they produce, in which calculation of failure rates can be used for those devices or systems. For equipment or systems that produce recently, the historical data of similar equipment or systems can serve as a useful estimate.

- Government and commercial failure rate data.

  The available handbooks of failure rate data for various equipment can be obtained from government and commercial sources. MIL-HDBK-217F, reliability prediction of electrical equipment, is a military standard that provides failure rate data for many military electronic components. Several failure rate data sources are available commercially that focus on commercial components, including some non-electronic components.

- Testing

  The most accurate source of data is to test samples of the actual devices or systems in order to generate failure data. This is often prohibitively expensive or impractical, so that the previous data sources are often used instead.

4.7. Failure distribution types

The different types of failure distribution are provided in **Table 2**. For an exponential failure distribution the hazard rate is a constant with respect to time (that is, the distribution is

Distributions			
Discrete		**Continuous**	
Binomial	Covered	Normal	Covered
Poisson	Covered	Exponential	Covered
Multinomial	Beyond the scope	Lognormal	Covered
		Weibull	Covered
		Extreme value	Beyond the scope

Table 2. Failure distribution types.

"memoryless"). For other distributions, such as a Weibull distribution or a log-normal distribution, the hazard function is not constant with respect to time. For some such as the deterministic distribution it is monotonic increasing (analogous to "wearing out"), for others such as the Pareto distribution it is monotonic decreasing (analogous to "burning in"), while for many it is not monotonic.

4.8. Derivations of failure rate equations for series and parallel systems

This section shows the derivations of the system failure rates for series and parallel configurations of constant failure rate components in Lambda Predict.

4.9. Series system failure rate equations

Consider a system consisting of n components in series. For this configuration, the system reliability, Rs, is given by [4]:

$$R_S = R_1 . R_2 . \dots . R_n \tag{9}$$

where R_1, R_2, …, Rn are the values of reliability for the n components. If the failure rates of the components are λ_1, λ_2,…, λ_n, then the system reliability is:

$$\begin{aligned} R_S &= e^{-\lambda_1 t} . e^{-\lambda_2 t} . \dots . e^{-\lambda_n t} \\ &= e^{-(\lambda_1+\lambda_2+\dots+\lambda_n)t} \end{aligned} \tag{10}$$

Therefore, the system reliability can be expressed in terms of the system failure rate, λ_S, as:

$$R_S = e^{-\lambda_s t} \tag{11}$$

Where $\lambda_s = \sum_{i=1}^{n} \lambda_i$ and λ_S is constant. Note that since the component failure rates are constant, the system failure rate is constant as well. In other words, the system failure rate at any mission time is equal to the steady-state failure rate when constant failure rate components are arranged in a series configuration. If the components have identical failure rates, λ_C, then:

$$\lambda_s = n\lambda_c \tag{12}$$

It should be pointed out that if n blocks with non-constant (i.e., time-dependent) failure rates are arranged in a series configuration, then the system failure rate has a similar equation to the one for constant failure rate blocks arranged in series and is given by:

$$\lambda_s(t) = \sum_{i=1}^{n} \lambda_i(t) \tag{13}$$

4.10. Parallel system failure rate equations

Consider a system with n identical constant failure rate components arranged in a simple parallel configuration. For this case, the system reliability equation is given by:

$$R_s = 1 - (1 - R_c)^n \tag{14}$$

where R_C is the reliability of each component. Substituting the expression for component reliability in terms of the constant component failure rate, λ_C, yields:

$$R_s = 1 - (1 - e^{-\lambda_c t})^n \tag{15}$$

Notice that this equation does not reduce to the form of a simple exponential distribution like for the case of a system of components arranged in series. In other words, the reliability of a system of constant failure rate components arranged in parallel cannot be modeled using a constant system failure rate model.

To find the failure rate of a system of n components in parallel, the relationship between the reliability function, the probability density function and the failure rate is employed. The failure rate is defined as the ratio between the probability density and reliability functions, or:

$$\lambda_s = \frac{f_s}{R_s} \tag{16}$$

Because the probability density function can be written in terms of the time derivative of the reliability function, the previous equation becomes:

$$\lambda_s = \frac{-\frac{dR_s}{dt}}{R_s} \tag{17}$$

The reliability of a system of n components in parallel is:

$$R_s = 1 - (1 - R_c)^n \tag{18}$$

and its time derivative is:

$$\frac{dR_s}{dt} = -n\,(1 - R_c)^{n-1}\frac{dR_c}{dt} \tag{19}$$

Substituting into the expression for the system failure rate yields:

$$\lambda_s = \frac{-n\,(1-R_c)^{n-1}\,\frac{dR_c}{dt}}{1-(1-R_c)^n} \tag{20}$$

For constant failure rate components, the system failure rate becomes:

$$\lambda_s = \frac{n\lambda_c e^{-\lambda_c t}(1-e^{-\lambda_c t})^{n-1}}{1-(1-e^{-\lambda_c t})^n} \tag{21}$$

Thus, the failure rate for identical constant failure rate components arranged in parallel is time-dependent. Taking the limit of the system failure rate as t approaches infinity leads to the following expression for the steady-state system failure rate:

$$\begin{aligned}\lambda_{s,steady\ state} &= \lim_{t\to\infty}\lambda_s \\ &= \lim_{t\to\infty}\frac{n\lambda_c e^{-\lambda_c t}(1-e^{-\lambda_c t})^{n-1}}{1-(1-e^{-\lambda_c t})^n} \\ &= n\lambda_c \lim_{t\to\infty}\frac{e^{-\lambda_c t}(1-e^{-\lambda_c t})^{n-1}}{1-(1-e^{-\lambda_c t})^n}\end{aligned} \tag{22}$$

Applying L'Hopital's rule one obtains:

$$\lambda_{s,steady\ state} = n\lambda_c \tag{23}$$

So the steady-state failure rate for a system of constant failure rate components in a simple parallel arrangement is the failure rate of a single component. It can be shown that for a k-out-of-n parallel configuration with identical components:

$$\lambda_{s,steady\ state} = n\lambda_c \tag{24}$$

Author details

Fatemeh Afsharnia

Address all correspondence to: afsharniaf@yahoo.com

Department of Biosystems Engineering, Ramin Agriculture and Natural Resources University of Khuzestan, Ahvaz, Khuzestan, Iran

References

[1] Billinton R, Allan RN. Reliability Evaluation of Engineering Systems (Concepts and Techniques). New York, London: Plenum Press; 1992. 453 pp

[2] Available from: http://www.reliabilityeducation.com/ReliabilityPredictionBasics.pdf

[3] Available from: http://www.mtl-inst.com

[4] Available from: http://www.weibull.com/hotwire/issue181/article181.htm

Chapter 8

General Perspectives on Seismic Retrofitting of Historical Masonry Structures

Baris Sayin, Baris Yildizlar, Cemil Akcay and
Tarik Serhat Bozkurt

Additional information is available at the end of the chapter

http://dx.doi.org/10.5772/intechopen.69439

Abstract

This chapter focuses on retrofitting of historical masonry structures from the point of seismic resistance based on failure analysis. In historical structures, restoration applications have become necessary because their life cycle of structural and nonstructural members is completed due to natural result of material structure, environmental conditions, and/or user errors. One of the most important intervention decisions in restoration stages carried out in historical buildings is known as retrofitted of the structure. The choice techniques of retrofitting of the structural members are becoming a very important issue in the scope of restoration of historical masonry structures belonging to the cultural heritage. Additionally, it should be decided to optimally preserve such buildings' original forms and to make interventions to increase the building's service life; in this regard, it is important to preserve the structures' historical identity and constructional value. Therefore, retrofitting applications have become essential to prevent the damage level and to have adequate level of structural strength in order to resist dynamic effects such as earthquakes. In this chapter, it is aimed to determine the main principles by using conventional and modern techniques within the scope of laboratory tests and numerical approaches in recovering the historical structures.

Keywords: historical structures, retrofitting, numerical approach, visual inspection

1. Introduction

With respect to ICOMOS 2001 document, a full understanding of the structural behavior and material characteristics is essential for any conservation and restoration project regarding the architectural heritage. It is suggested that the work of analysis and evaluation should be

INTECH
open science | open minds

© 2017 The Author(s). Licensee InTech. This chapter is distributed under the terms of the Creative Commons Attribution License (http://creativecommons.org/licenses/by/3.0), which permits unrestricted use, distribution, and reproduction in any medium, provided the original work is properly cited.

done with multidisciplinary approach, such as earthquake specialists, architects, engineers, and art historians. Additionally, it is considered necessary for these specialists to have common knowledge on the subject of conserving and upgrading or strengthening the historical structures [1].

Masonry constructions in many countries worldwide are characterized by inadequate resistance to earthquakes loading. The use of appropriate techniques for retrofitting of historical masonry structures should be made by referencing to additional structural system and members as well as repair and retrofitting on period after built of the mentioned structures, and this fact is evaluated in terms of the protection of the original identity as well as the cultural value of the structures [2].

The preservation of the original form in the retrofitting applications of historical structures is taken into consideration and thus transferring building's historical identity to future generations may be possible. Besides that, the need for refunctioning of the structure, improvement of repair and strengthening techniques over time, technical specifications in force, and collaboration of experts of different disciplines leads to the development of new solutions in the restoration applications.

Repair and strengthening applications are commonly applied as part of restoration works of historical masonry structures for recent years [3–21]. Until 1980, the applied practices consist of removing floor arch or timber floor by preventing load-bearing masonry wall, building a secondary load-bearing system apart from the present bearing system by supplying a RC load-bearing system inside large internal spaces, and ruining architectural structure by joining columns and beams inside spaces.

In this study, seismic retrofitting applications in historical masonry structures are presented within the scope of laboratory tests and numerical analyses based on a cross-disciplinary study of civil engineering and architecture.

2. Failure analysis

Structural failures and their investigation have become an active field of professional practice in which experts are retained to investigate the causes of failures, as well as to provide technical assistance to know the root cause. Failure need not always mean a structural collapses. It can make a structure deficient or dysfunctional in usage. It may even cause secondary adverse effects: (i) Safety failure (injury, death, or even risk to people), (ii) functional failure (compromise of intended usage), and (iii) ancillary failure (adverse effect on schedules, cost, or use) [22–24].

Failure analysis and prevention are significant parameters to all of the engineering disciplines. The materials engineer often plays a head role in the analysis of failures, whether a component or product fails in service or if failure occurs in manufacturing or during production processing. In all cases, one must determine the cause of failure to inhibit future occurrence, and/or to increase the performance of the structure [25]. Generally, procedure for failure analysis includes: (i) collection of data and samples, (ii) preliminary examination, (iii) nondestructive inspection, (iv) mechanical testing, (v) selection and preservation of damaged surfaces, (vi) macroscopic and microscopic examination, (vii) preparation and examination of metallographic sections, (viii) damage classification, and (ix) report writing. In this chapter, the mentioned procedures are considered.

3. Intervention strategies

One of the most important intervention decisions in restoration carried out in historical structures is retrofitting of the structure. Article 10 of the Venice Charter (1964) remarks that "Where traditional techniques prove inadequate, the consolidation of a monument can be achieved by the use of any modern technique for conservation and construction, the efficacy of which has been proven by scientific data and experience" [1].

Retrofitting applications in historical structures are performed in accordance with the intervention decisions based on conventional and modern techniques. The intervention solutions must rely on cost-benefit analyses and take into account their socioeconomic impact on society. An obvious requirement is to minimize as much as possible the disturbance of the owners of the building during the building rehabilitation. The financial resources available decisively influence the intervention solutions for the particular purposes, including labor work capacity, equipment, materials, duration of the work, etc. It is also compulsory to have alternative strategies for intervention and to evaluate the decrease in building vulnerability with various strategies [26].

The intervention strategy and the intervention techniques must take into account the following criteria: (i) seismic hazard level at the construction site, (ii) characteristics of the building's intended use (architectural constrains, original occupancy of the building, building structure, technical equipments within the building, etc.), (iii) building safety as a response to daily activities, mainly related to the seismic safety (structural vulnerability, vulnerability of nonstructural elements, appliances or/and equipments, building exposure or value, etc.), (iv) required level for building performance (life safeguarding, immediate occupancy after earthquake, preventing building collapse, etc.), (v) economic criteria, including insured and reinsured value of the building, and (vi) technological capability available at the site.

4. Retrofitting principles

As a rule, it is worth stating the preliminary consideration before the description of the retrofitting techniques. Once the technique is proposed, it should be experimentally studied to understand the best application procedure and also its effectiveness. Even so, no general repair and strengthening method exists due to variability of historical masonry structures in materials and construction techniques. On the contrary, methods are mostly determined according to different masonry characteristics.

Appropriate site and laboratory investigations should always be carried out before reinforcement application through: (i) an accurate geometrical survey of the structure's morphology; (ii) characterization of the constituent materials, e.g., stones and of the mortars in masonry structures; (iii) survey of the physical and mechanical decay; and (iv) crack pattern survey. Based on these considerations, it is possible to choose between several systems of retrofitting and application techniques.

Three important preservation principles should be kept in mind when undertaking seismic retrofit projects: [26]

- Historical materials should be preserved at the largest extent possible and not replaced wholly with other new materials in the process of seismic strengthening.
- New seismic retrofit systems, whether hidden or exposed, should respect and should be compatible to the architectural and structural integrity of the historical structure.
- Seismic work should be "reversible" to the largest extent possible to allow future removal for the use of future improved systems.

While seismic upgrading work is often more permanent than reversible, care must be taken to preserve historical materials to the historical appearance of the building.

In addition to the above principles, the general aim of structure conservation is to preserve the cultural significance of the place where the building is culturally aggregated. Places of cultural significance should be safeguarded and not left at risk in a vulnerable state.

The different intervention works are defined in various manners and documents from various countries, as well as in general literature. The relationship between the stages for seismic rehabilitation of historical buildings is presented in **Figure 1**.

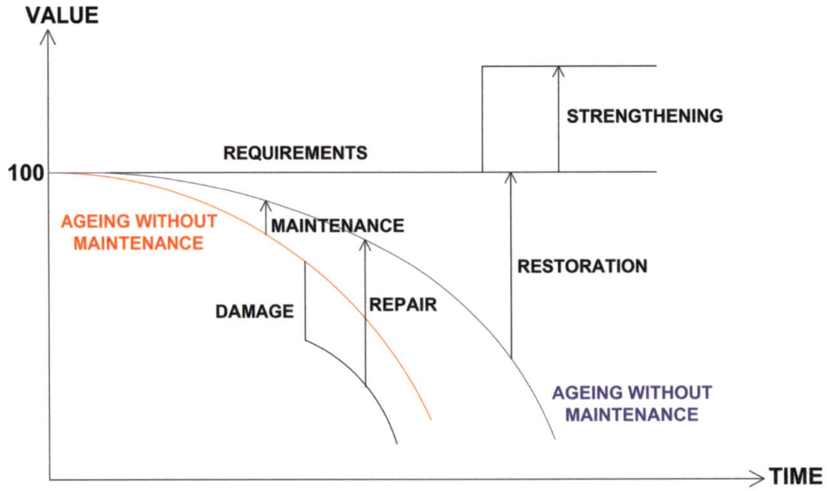

Figure 1. Value alteration diagram of a building [26].

The seismic retrofitting methods can be classified as follows, as far as their target is concerned:

a. In case the target is the increase of stiffness and structural strength, the most effective method is the addition of walls in the existing frames. What follows is the method of adding truss bracings, the method of adding walls as an extension of existing columns, and the use of composite materials.

b. In case the target is the increase of plasticity, the method that is recommended is the application of jacketing on a number of selected columns, as well as the use of composite materials.

c. In case the target is the simultaneous increase of strength, stiffness, and plasticity of the structure, any seismic strengthening method can be used taking into consideration the required degree of increasing each of the above-mentioned characteristics. In case that the necessary increases are very high for all the three characteristics, the addition of new vertical members is generally inevitable.

5. Structural evaluation of masonry structures

5.1. Historical process of the structures

Within the scope of examined masonry structures, in order to determine structural changing or the construction time of an examined structure, it should be investigated whether or not the structure is in place using ancient maps. In the stage of the restoration project, historical photos and old maps are utilized in collecting data and documents on the building facade; therefore, information on floor height, storey count, roof patio, roof form of the structure, relationship with the neighboring structures, and location of the windows is obtained.

5.2. Existing status of the structures

The current state of an examined structure must be defined within the context of the structural and nonstructural members and the position in the plane must be given. Three-dimensional (3D) laser scanning method should be used to determine the geometric dimensions and structural damage of the structure in detail. Then, the structural supplements in the historical process of examined structure must be presented in detail. Thus, the differences between the first period and the current situation of an examined structure will be clarified.

5.3. Field study

Samples are gathered from various points of a structure during laboratory test phase of the study and after then analyses are performed. Thereby, axial compressive strength of the masonry walls, shear strength, bulk density, and thick- and thin-plaster compounds are derived. Moreover, mechanical characteristics of bricks and pointing fillings are determined.

5.3.1. Shear strength of walls in situ

Shear strength index of brick-mortar in masonry walls are defined based upon shear tests which are performed in different points in the structure in relation to standard. Samples are compressive strength and bulk density tests are performed on such samples—taken representing the brick-mortar-brick system—in the laboratory.

5.3.2. Mechanical and physical properties of mortars and brick samples

Samples from the solid brick blocks and pointing filling mortars are gathered from an examined structure. Single-axis compressive strength and bulk density tests are performed on brick blocks. For the purpose of determining the dead weight of the wall, density of filling mortars between the bricks is considered.

5.3.3. Elastic modulus of walls

In consequence of the tests performed on an examined structure, in parallel with the data and previous experiences acquired from other masonry structures having comparable historical process, construction conditions, and materials, it is decided that the elastic modulus, taking into consideration the condition of the existing walls, can be used on the structural model.

5.4. Laboratory analysis

As a laboratory study, physical test and chemical analysis are carried out on samples taken from a historical building. Subsequently, the proposal mixtures of plaster and mortar for restoration are determined according to the laboratory test results. The locations of samples from the masonry ruin are determined. Properties and detailed visual definition of the provided samples are obtained. Then, samples are taken from different points of the examined structure during lab test stage of the study and analyses are carried out. The samples are exposed to chemical, physical, and mineralogical tests, such as the ratio of binders, aggregates, and mineralogical compositions, and textural properties of the mortar samples are analyzed, and the cause for deteriorations is determined by visual analysis, spot tests analysis, ignition loss analysis, reaction with acid, and petrography. Before starting the analyzes, it is examined that the texture, color, state (integrity) of the samples; color, size, approximate quantities, apparent organic additives and type of aggregates in the contents of the samples. After that, the size distribution, particle properties of acid insoluble materials, and protein/ oil contents are determined. Samples taken are examined by content analysis within the scope of related standards along with visual detailing. Consequently, axial compressive strength of the masonry walls, shear strength, bulk density, and thick and thin plaster compounds are obtained. Also, mechanical characteristics of bricks and pointing fillings are obtained.

5.5. Numerical analysis

In the numerical analysis carried out with general purpose, finite element analysis software, period values, modal participation mass ratios, and total modal participation mass ratios corresponding to the 12 modes are calculated for an examined structure in modal analysis. The period values are found out for x- and y-directions. For each earthquake direction in the regulations [27], shear stresses on the walls are attained by rating the horizontal force coming to the horizontal load-bearing members to horizontal section area of masonry walls. After that, calculated values are compared to the experimental shear stress ones. Shear stresses of all walls in both directions of a masonry structure are determined to be adequate to cover the shear stresses under earthquake effects.

5.5.1. Structural model

FE analysis of a multistorey masonry structure under seismic loading is performed. Wall shear strength to be used in numerical analyses is determined individually for each of the storey. The structural performances are determined based on the existing building's load-bearing data obtained by using linear elastic analyses. According to the analyses, principles defined as per related country's local provisions and standards, all analyses are carried out using a general purpose FE software [28, 29]. In terms of evaluation of the existing structures, the structure is analyzed considering four different building performance levels for the existing or strengthened buildings, in addition to various level earthquake definitions. The performance levels are specified as: (a) operational, (b) immediate occupancy, (c) life safety, and (d) collapse prevention or near collapse (**Figure 2**).

Building performance can be defined qualitatively from the point of: (i) safety afforded building occupants, during and after an earthquake; (ii) cost and feasibility of restoring the building to preearthquake circumstances; (iii) length of time the building is removed from service to conduct repairs; and (iv) economical, architectural, or historical effects on the community in general. These performance properties will be squarely related to the extent of damage sustained by the building during an earthquake.

In the numerical analyses performed under a described earthquake design, it is specified that the existing system meets the conditions set forth for life safety performance level. For an earthquake with 10% recovery possibility in 50 years, it is determined that the shear strength of all the walls in both directions of the masonry building is ordinarily adequate to

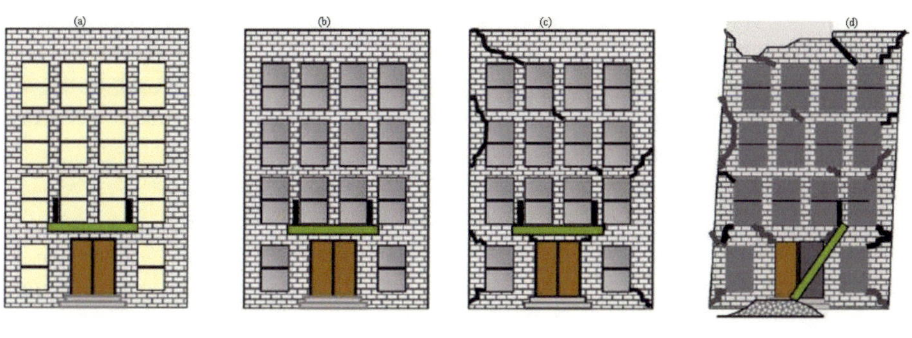

Figure 2. Levels of building performance: (a) operational, (b) immediate occupancy, (c) life-safety, and (d) collapse prevention [30].

meet the shear forces occurring under the subject to earthquake effects. In that case, it can be underlined that a structure meets that "immediate usage performance level"; however, due to insufficiencies not meeting a masonry structure definitions of the structure and partially determined irregularities, "life safety performance level" should be used to evaluate, rather than "immediate usage performance level."

Building performance levels typically comprise a structural performance level that describes the limiting damage state of the structural systems, plus a nonstructural performance level that describes the limiting damage state of the nonstructural systems and components. A structure examined by using mode combination method is calculated under earthquake loads; all acceptances and obtained values are determined in accordance with the Turkish Earthquake Regulation. A three-dimensional model of an examined structure is prepared. In the prepared model walls and slabs that are horizontal structural members are modeled as shell element in the element library of the FE analysis software. Mechanical properties of the structural members in the model are transferred to the structural system via the data obtained from the experimental studies mentioned above.

5.5.2. Analysis of structures

5.5.2.1. Compliance with earthquake regulation

In determining the structure's performance, controls are made based on local or international earthquake regulations [26]. A three-dimensional linear finite element analysis of examined multistorey masonry structure under seismic loading is carried out.

The following checks should be evaluated in accordance with seismic regulation:

- Control of the total maximum length of the masonry walls
- Nonsupported length between the vertically connected masonry wall axes to the plane of any masonry wall is checked in according to earthquake zone
- Limit conditions in door and window gaps in the walls
- The height of each storey should be checked in masonry buildings

Based on the regulation, supported walls in masonry structures should be as regular and arranged as a symmetrically or close to symmetric, and construction of local basement should be avoided. The structure is examined considering these criteria and it is determined whether the structure is regular or not on the plan.

5.5.2.2. Modal analysis

Twelve modes are calculated for the building in modal analysis and period values corresponding to the modes are presented. Participation mass ratio and period values are examined and the vertical and horizontal period values are obtained. Data presented demonstrate that which direction is more rigid than the other direction.

5.5.2.3. Checking of shear stress in walls

Shear stresses on the walls are obtained by proportioning the horizontal force coming to the horizontal load-bearing members to horizontal section area of masonry wall for each earthquake direction in the earthquake provision. Calculated values are compared to the experimental shear stress values. Shear stresses are calculated individually for each storey and wall across building height. Total shear force is calculated for FE analysis for each direction by proportioning the total section area in that direction, and principal stress comparisons are made on the shell elements together with shear stress for each storey.

Storey shear forces obtained from analysis are compared to the forces that are experimentally obtained and shear stress checking is carried out in the structure. Shear performance on x- and y-directions is presented. Shear strength of all the walls in both directions of the masonry building is determined to be adequate to cover the shear forces under earthquake effects. Also, it is checked whether the checking meet the rules for the structure geometry and design.

6. The retrofitting applications of masonry structures

In the retrofitting stage of the structures, plaster analyses and mechanical tests are conducted on the samples taken from the examined structure. After that, the structure's existing 3D computer model is prepared and members that are inadequate in terms of strength are determined. Within the scope of the analysis, it is found that the structure can be converted to the original form of the structure. In this regard, featureless additions, which are not related to the original structure, are removed and restoration applications are carried out. In the scope of restoration applications, structural cracks on the walls are repaired using the injection method. Moreover, jack-arched floor, exterior facade walls, interior walls, and door/window gaps using different techniques are strengthened.

As a cross-disciplinary study must be carried out in the strengthening stage of the historical building, improved or changed details are manufactured with the approval of engineering and architecture disciplines. Retrofitting applications include: (i) walls, (ii) jack-arched floors (volta slabs), (iii) door and window gaps, and (iv) foundations. In this section, retrofitting applications illustrating the effectiveness of the described method will be presented. It should be noted that the presented applications are for example only and are not intended to be exhaustive.

6.1. Retrofitting of masonry walls

6.1.1. Retrofitting of cracks in masonry walls using injection technique

In an examined structure, it is decided that masonry walls will be strengthened removing plastering on masonry walls, detection of structural cracks and strengthening using the method of injection in accordance with supervisor decisions regarding strengthening masonry walls.

It is aimed to increase the bearing capacity of masonry walls using injection method, as filling the structural slots inside the masonry walls with lime binder injection mortar.

Following completing plaster removal on all masonry walls, structural cracks are determined. Repair mixture ratios are determined based on the gathered samples, and the mixtures are prepared. The mixture is applied to the cracks by means of manual injection. Integration is ensured using steel clamp members in larger cracks and anchorage elements on the walls are anchored via epoxy.

As it takes time to settle the injection compound into the wall, application is repeated periodically. The compound inside the pipe is penetrated into the interior walls in time, the pipes are continuously injected, and this process is continued until the gaps are filled. **Figure 3** shows possible injection application in the walls.

6.1.2. Retrofitting of masonry walls in-plane using carbon fiber wrap

Carbon fiber wrap can be used in tensile stress areas on wall members to strength dynamic effects, e.g., earthquake. First, applied carbon fiber locations are specified on the facades. Carbon fiber wrap is positioned between window gaps and storey levels and also disguised under plaster layer. **Figure 4** shows possible carbon fiber application in the walls.

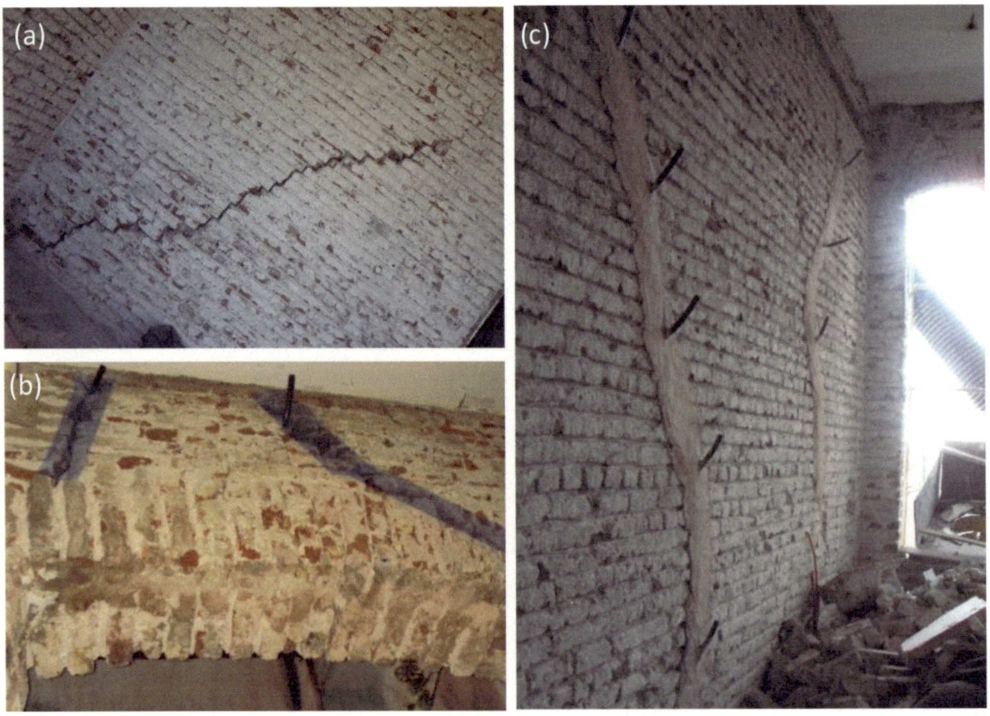

Figure 3. Injection application in the walls after plaster removal: (a) structural cracks, (b) preparation of wall surface for injection, and (c) injection application by manual technique.

Figure 4. Carbon fiber wrap applications on masonry walls (a and b).

6.1.3. Retrofitting of internal walls using rebar grids technique

After having completed the existing plaster removal from the vertical load-carrying members, internal wall strengthening is applied. In this respect, anchorage holes per square meter are drilled on masonry walls and filled with epoxy (**Figure 5**).

Figure 5. Retrofitting of the internal walls after plaster removal: (a–c) rebar grid technique and (d) application of thick plaster.

6.2. Retrofitting of jack-arched floors

Floor system strengthening consists of two separate parts: (i) Strengthening without removing jack-arched floors, which are positioned on the hallways and assumed as the original ones and (ii) applications of steel beam system installation is applied with the aim of lighting the weight of the building as removing featureless RC additions. **Figure 6** presents a similar application about floor strengthening.

Figure 6. Retrofitting of jack arched slabs: (a) removing top layers of the slab, (b) welding of shear rebar over I-profile beam, (c) steel tie, (d) connections of stainless steel ties between the masonry walls, (e) anchorage of stainless steel, and (f) topping concrete application on slab.

6.3. Enhancement of lateral stability of door and window gaps

The locations and dimensions of original windows and doors are based upon a restoration process, and all infilled the windows and doors are opened on the facades in the current situation. Then, it can be strengthened the window- and door-gaps with steel plate to increase the strength in order to resist dynamic effects according to results of 3D analysis (**Figure 7**).

Figure 7. Retrofitting of door and window gaps using steel plates (a and b).

6.4. Retrofitting of foundations

For foundation strengthening, it is managed that all masonry walls are to move together by tying them with each other with bond beams and RC foundations and also aimed to transfer the load of the building to foundation system once and later to the ground (**Figure 8**).

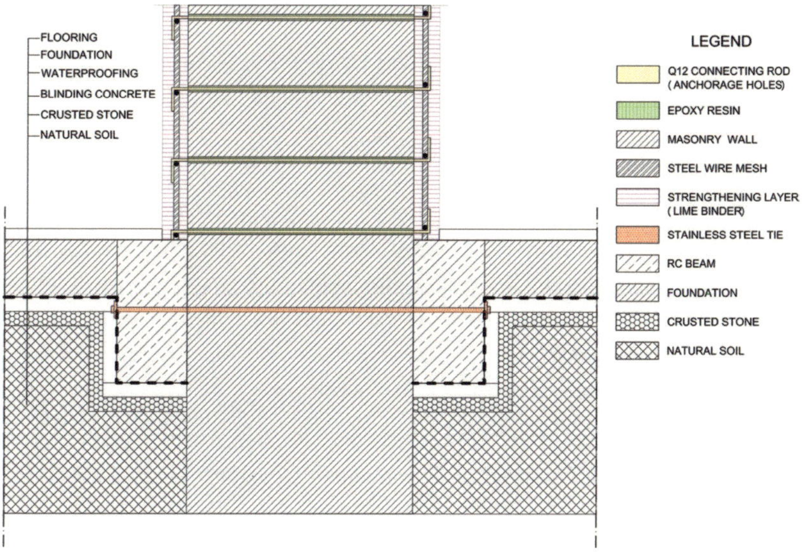

Figure 8. Retrofitting of foundations in masonry structures.

7. Conclusion

The chapter provides a general overview on the retrofitting of historical masonry structures which have critical issues in terms of seismic resistant. It is concluded that the retrofitting process is suitable as a practical tool for retrofitting applications. Retrofitting techniques of masonry structures are highly effected from the scientific and technological advances. Through the last decades, many retrofitting techniques have been proposed and applied to the structures. Accuracy of retrofitting methods depends mainly upon analyses of examined structures and classification techniques. The efficiency of the retrofitting for historical masonry structures is directly related to the suitability of the used methods or techniques with retrofitting principles. Moreover, needless to say that in addition to the retrofitting of the structures, maintenance and repair of the structures also plays a major role in its service life.

Acknowledgements

The authors like to thank the Department of Construction and Technical Affairs, Istanbul University Rectorate, for valuable supports.

Author details

Baris Sayin[1]*, Baris Yildizlar[1], Cemil Akcay[1] and Tarik Serhat Bozkurt[2]

*Address all correspondence to: barsayin@istanbul.edu.tr

1 Department of Civil Engineering, Istanbul University, Istanbul, Turkey

2 Department of Architecture, Istanbul Technical University, Istanbul, Turkey

References

[1] ICOMOS (International Council on Monuments and Sites) Recommendations for the Analysis, Conservation and Structural Restoration of Architectural Heritage, Paris, International Scientific Committee for Analysis and Restoration of Structures of Architectural Heritage, 2001

[2] Corradi M, Osofero AI, Borri A, Castori G. "Strengthening of historic masonry structures with composite materials". Handbook of Research on Seismic Assessment and Rehabilitation of Historic Structures (2 Volumes) Chapter 8, Eds. Asteris PG, Plevris V. 2015; p. 257-292

[3] Bozkurt TS, Sayin B, Karakas AS, Akcay C, Yildizlar B. The preserving and improvement of historical structures based on qualified a RC structure: A case study.

CACMSISTANBUL 2015: International Conference on Advances in Composite Materials and Structures, Chair: Professor Antonio Ferreira, Faculty of Engineering, University of Porto, Portugal, Book Abstracts: p. 87-88, April 13-15, 2015, Istanbul, Turkey

[4] Bozkurt TS, Sayin B, Akcay C, Yildizlar B, Karacay N. Restoration of the historical masonry structures based on laboratory experiments. Journal of Building Engineering. 2016;**7**:343-360. DOI: 10.1016/j.jobe.2016.07.010

[5] Akcay C, Bozkurt TS, Sayin B, Yildizlar B. Seismic retrofitting of the historical masonry structures using numerical approach. Construction and Building Materials. 2016;**113**:752-763. DOI: 10.1016/j.conbuildmat.2016.03.121

[6] Sayin B, Bozkurt TS, Yildizlar B, Akcay C. The consolidation of historical masonry structures: A case study. In: ACE 2016: 12th International Congress on Advances in Civil Engineering; 21-23 September 2016; Boğaziçi University, Istanbul, Turkey

[7] Sayin B, Akçay C, Yıldızlar B, Bilir T, Bozkurt TS. Restoration approach to improve sustainability and longevity in existing historical structures. In: SCMT4: 4th International Conference in Sustainable Construction Materials and Technologies; 7-11 August 2016, Las Vegas, USA.

[8] Karakaş AS, Bozkurt TS, Sayın B, Akcay C, Yıldızlar B. The restoration applications on historical structures: A case study, ISITES 2015: 3rd International Symposium on Innovative Technologies in Engineering and Science, Universidad Politécnica de Valencia; Spain, 3-5 June, 2015.

[9] Asteris PG, Chronopoulos MP, Chrysostomou CZ, Varum H, Plevris V, Kyriakides N, Silva V. Seismic vulnerability assessment of historical masonry structural systems. Engineering Structures. 2014;**62-63**:118-134. DOI: 10.1016/j.engstruct.2014.01.031

[10] Ahunbay Z. Conservation and Restoration of Historical Environment (Tarihi çevre koruma ve restorasyon). YEM Publishing; 2004. p. 134. ISBN: 975-7438-38-3, Istanbul, Turkey

[11] Chmielewski R, Kruszka L. Application of selected modern technology systems to strengthen the damaged masonry dome of historical St. Anna's Church in Wilanów (Poland). Case Studies in Construction Materials. 2015;**3**:92-101. DOI: 10.1016/j.cscm.2015.08.001

[12] Valluzzi MR, Bondì A, Porto F, Franchetti P, Modena C. Structural investigations and analyses for the conservation of the 'Arsenale' of Venice. Journal of Cultural Heritage. 2002:**3**:65-71. DOI: 10.1016/S1296-2074(02)01161-5

[13] Bednarz LJ, Jasienko J, Rutkowski M, Nowak TP. Strengthening and long- term monitoring of the structure of an historical church presbytery. Engineering Structures. 2014;**81**:62-75. DOI: 10.1016/j.engstruct.2014.09.028

[14] Barbieri G, Biolzi L, Bocciarelli M, Fregonese L, Frigeri A. Assessing the seismic vulnerability of a historical building. Engineering Structures. 2013;**57**:523-535. DOI: 10.1016/j.engstruct.2013.09.045

[15] Ascione L, Feo L, Fraternali F. Load carrying capacity of 2D FRP/strengthened masonry structures. Composites. Part B. Engineering. 2005;**36**(8):619-626. DOI: 10.1016/j.compositesb.2004.12.004

[16] Borri A, Castori G, Corradi M. Intrados strengthening of brick masonry arches with composite materials. Journal of Composites, Part B. 2011;**42**(5):1164-1172. DOI: 10.1016/j.compositesb.2011.03.005

[17] Corradi M, Borri A, Vignoli A. Strengthening techniques tested on masonry structures struck by the Umbria–Marche earthquake of 1997-1998. Construction and Building Materials. 2002;**16**(4):229-239. DOI: 10.1016/S0950-0618(02)00014-4

[18] Corradi M, Borri A, Vignoli A. Experimental evaluation of in-plane shear behavior of masonry walls retrofitted using conventional and innovative methods. Masonry International. 2008;**21**(1):29-42. ISSN 0950-2289

[19] Corradi M, Tedeschi C, Binda L, Borri A. Experimental evaluation of shear and compression strength of masonry wall before and after reinforcement: Deep repointing. Construction and Building Materials. 2008;**22**(4):463-472. DOI: 10.1016/j.conbuildmat.2006.11.021

[20] Krevaikas TD, Triantafillou TC. Masonry confinement with fiber-reinforced polymers. Journal of Composites for Construction. 2005;**9**(2):128-135. DOI: 10.1061/(ASCE)1090-0268(2005)9:2(128)

[21] Triantafillou TC. Strengthening of masonry structures using epoxy-bonded FRP laminates. Journal of Composites for Construction. 1998;**2**(2):96-104. DOI: 10.1061/(ASCE)1090-0268(1998)2:2(96)

[22] Ratay RT. Professional practice of forensic structural engineering. Structure Magazine, July 2007. pp. 50-53

[23] Krishnamurthy N. Forensic engineering in structural design and construction. In: CD Preprints of Structural Engineers World Congress; 2-7 November, 2007, Bangalore, India, p. 2.

[24] Pathan KM, Ali SW, Zubair S, Najmi AH. A forensic view to structures' failure analysis. SSRG International Journal of Civil Engineering. 2015;**2**:26-32

[25] Pizzo PP. Exploring Materials Engineering, Failure Analysis. Available from: engr.sjsu.edu/WofMatE/FailureAnaly.htm [Accessed: January 2017]

[26] Lungu D, Arion C. 2005, Intervention strategies. Prohitech project: Earthquake protection of historical buildings by reversible mixed technologies. Chapter 5. pp. 1-85, FP6-2002-INCOMPC-101.

[27] TEC 2007. Turkish Earthquake Code for Buildings, Specification for Buildings to be Built in Earthquake areas. Ankara, Turkey: Ministry of Public Works and Resettlement; 2007

[28] TS 498. The Loads and Loading Cases Due to use and Occupancy in Residential and Public Buildings. Turkish Standards Institute; 1997, Ankara, Turkey.

[29] TS 500. Requirements for Design and Construction of Reinforced Concrete Structures. Ankara, Turkey; Turkish Standards Institute; 2000

[30] FEMA 389. Primer for design professionals communicating with owners and managers of new buildings on earthquake risk. Risk Management Series. January 2004, Federal Emergency Management Agency USA

Chapter 9

Common Case Studies of Marine Structural Failures

Goran Vukelić and Goran Vizentin

Additional information is available at the end of the chapter

http://dx.doi.org/10.5772/intechopen.72789

Abstract

Marine structures are designed with a requirement to have reasonably long and safe operational life with a risk of catastrophic failures reduced to the minimum. Still, in a constant wish for reduced weight structures that can withstand increased loads, failures occur due to one or several following causes: excessive force and/or temperature induced elastic deformation, yielding, fatigue, corrosion, creep, etc. Therefore, it is important to identify threats affecting the integrity of marine structures. In order to understand the causes of failures, structure's load response, failure process, possible consequences and methods to cope with and prevent failures, probably the most suitable way would be reviewing case studies of common failures. Roughly, marine structural failures can be divided into structural failures of ships, propulsion system failures, offshore structural failure, and marine equipment failures. This book chapter will provide an overview of such failures taking into account failure mechanisms, tools used for failure analysis and critical review of possible improvements in failure analysis techniques.

Keywords: marine structures, failure analysis, fracture, fatigue, failure

1. Introduction

Marine structures must comply with such design requirements that the probability of failures or stability loss of parts and/or complete structures is reduced to minimum. Studies and analysis of marine structural failures had shown that a significant percentage of failures were a consequence of inadequate design due to lack of operational considerations, incomplete structural elements evaluations, and incorrect use of calculation methods.

Structural safety level is determined during design process by defining specific structural elements, material properties, and functional requirements based on the expected lifetime of the structure, ramifications of eventual failures and costs of failures. Time dependency

INTECH
open science | open minds

© 2017 The Author(s). Licensee InTech. This chapter is distributed under the terms of the Creative Commons Attribution License (http://creativecommons.org/licenses/by/3.0), which permits unrestricted use, distribution, and reproduction in any medium, provided the original work is properly cited.

of strength and loads has to be taken into account because the strength of a structure will decrease with time while the load is varying through the lifetime of the structure.

Successful material selection process implies reconciling requirements like suitable strength of material, sufficient level of rigidity, appropriate heat resistance, etc. Structures that are susceptible to crack growth need to be made of materials selected on the basis of fracture mechanics parameters. Fracture mechanics parameters that define material resistance to crack propagation are usually determined through experimental research, but nowadays some of the experiments can be successfully substituted with numerical analysis. Material fracture behavior is usually estimated using some of the well-established fracture parameters, like stress intensity factor (*K*), *J*-integral or crack tip opening displacement (CTOD). Besides that, fatigue limit has to be taken into account, also. It has become customary to perform an optimal fatigue design analysis as an integral part of design calculations. Such analyses are also largely based on data and procedures developed from experimental and empirical research.

Marine structural failures can be divided into three main groups: failures of ships, offshore structures, and marine equipment. This book chapter will provide an overview of most common case studies of such failures. Further, failure mechanisms will be emphasized and tools used for failure analysis outlined. Possible improvements in failure analysis techniques are discussed in the end of the chapter.

2. Common case studies

2.1. Ship structural failures

Maybe the most notable case of ship failures are failures of Liberty ships in the early 1940s. These failures gave a serious boost in the development of fracture mechanics. Ships, mass produced in assembly line style out of prefabricated sections as an all-welded construction, exhibited nearly 1500 cases of brittle fractures with 12 ships breaking in half. The results of failure investigation had shown that inadequate grade of steel allowed for brittle fracture at low temperatures. Further, rectangular hull openings, such as hatch square corners, that coincided with a welded seam acted as stress concentrations points and crack origins [1].

There has been a considerable amount of failures in recent times, also. For instance, structural failure of container ship MOL Comfort [2, 3] in 2013. A yearlong failure investigation concentrated on finding the possibility of fracture occurrence and structural safety level. Results had shown that the hull fracture originated from the bottom butt joint in the midship part. A possibility that the load's upper limit exceeded strength's lower limit was also estimated using probabilistic approach. Furthermore, safety inspections of the MOL Comfort sister ships have shown buckling deformations (concave and convex) of the bottom shell plating of up to 20 mm (4 mm allowable) in height observed near the center line. Finally, a numerical analysis of the ship hull taking the load history into account was done. After the investigation, it was concluded that the load of the vertical bending moment probably exceeded the hull girder ultimate strength when the deviations of the uncertainty factors are taken into account, which caused the bottom shell plates to buckle due to excessive load. The reduction of breadth of bottom shell plate between girders increased the stress in the girder which yielded in the lower part resulting in the collapse occurs in the middle part of the ship, at the bottom, near the center line.

Bilge keels structures are used to enhance the transverse stability of ships. Cracks have been noticed in various ships in the internal structure of the bilge keels and on the connecting points to the ship's hull. Failure analysis of the damage can identify the causes of failure and the analysis results serve as basis for design improvements. It has been shown, both theoretically and applying FEM analysis, that the failure locations in bilge keels structures occur in the stress concentration regions that are present due to the structure geometry commonly used, therefore new structural elements are proposed that significantly reduce the possibility of failure occurrence [4].

Corrosively aggressive cargo (acids, alkalis, etc.) can represent a danger to the integrity of ship structures. In the case of the "Stolt Rotterdam" freighter, which sank during the cargo loading in the port, the investigation (visual, macrofractographic, and chemical) following the sinking has shown that the residue valve has cracked due to a design-specific stress (stiffer main valve was missing), thus causing a leak of the acid that accelerated the corrosion process of the floor panels in the area of the leak. Also, the valve gaskets were made of a material not resistant to acid which also contributed to the speed of the leak [5].

Marine engines and propellers produce dynamic loads on their supportive structures which can lead to fatigue failures. One of the most stressed components of the engine structure is the bearing bushing foundation. A state-of-the-art design procedure for the bearing girders is comprised of essential procedures such as bearing loads determination, stresses calculation, and the bearing girder fatigue strength assessment [6]. The fatigue and structural durability analysis is conducted for multi-axial stresses and opens the possibility to construct lightweight engines.

2.2. Propulsion system failures

The propulsion system has a pivotal role on ships. A typical marine propulsion system is comprised of main engine, driving device, marine shaft, and propeller. Most of the failures occur on the propulsion shaft that is subjected to various types of loading during operation (torque moment, bending moment, axial thrust force, and transversal loads). The operating environment of the propulsion system is characterized by significant changes in temperatures and humidity, aggressive atmosphere, long-lasting interrupted operating time, and variations in load amplitudes. The risk of failures of the propulsion system additionally increases with the severity of sea and weather conditions as they have a direct effect on the dynamics of the load variation. All of the above has direct influence on fatigue behavior and life time of the propulsion shaft.

Shaft keys are recognized as a potential origin of growing cracks. The geometry of the ends of keyways represents a stress concentration factor in the cases of torque transmission through shaft keys for dynamic vibrational loads. Faulty machining of shaft key elements (key groove, keyway, and key) geometry, inadequate run out radii, or material imperfection can be root causes of torsional fatigue failure in shaft keys. The characteristic torsional failure indicator is the crack pattern that initiates at the end of the keyway and propagates in a 45° rotational direction in a helical shape. Also, interaction between engine body and hull must be taken into account, especially thermal loads that can affect the integrity of shafts and can be successfully solved numerically [7], **Figure 1**.

A case study [8] has shown that inadequate torsional vibration calculation parameters (shaft elements stiffness and damping, natural frequencies, safety factors) and a subsequent poor design of the shaft's keyway cause failures. In this case a root cause analysis was done by the analytical stress calculation process MIL G 17859D and VDI 3822 standards. A FEM model was used in order to verify the existing fracture characteristics and causes.

An alternative to shaft key joints are spline joints, which are press fitted to other shaft elements. Analysis of spline joint failure [9] shows that the press fitting of the joining elements can cause surface deformation which in turn causes surface cracks formation. Cracks usually start on the spline teeth at the shaft junction zone. Torsional fatigue caused by fluctuating stress promotes crack growth and propagation. Inhomogeneity of the shaft material can additionally assist crack propagation. In this case, visual and macroscopic inspection was performed, followed by material chemical analysis, hardness measurement, optical, and scanning electron microscope (SEM) microstructure analysis with X-ray dispersive analysis of particles under the SEM.

Bolted connections are used in collar coupling of shaft elements and in propeller blades connections. The changes of rotation direction of the shaft results in torque moment overloading

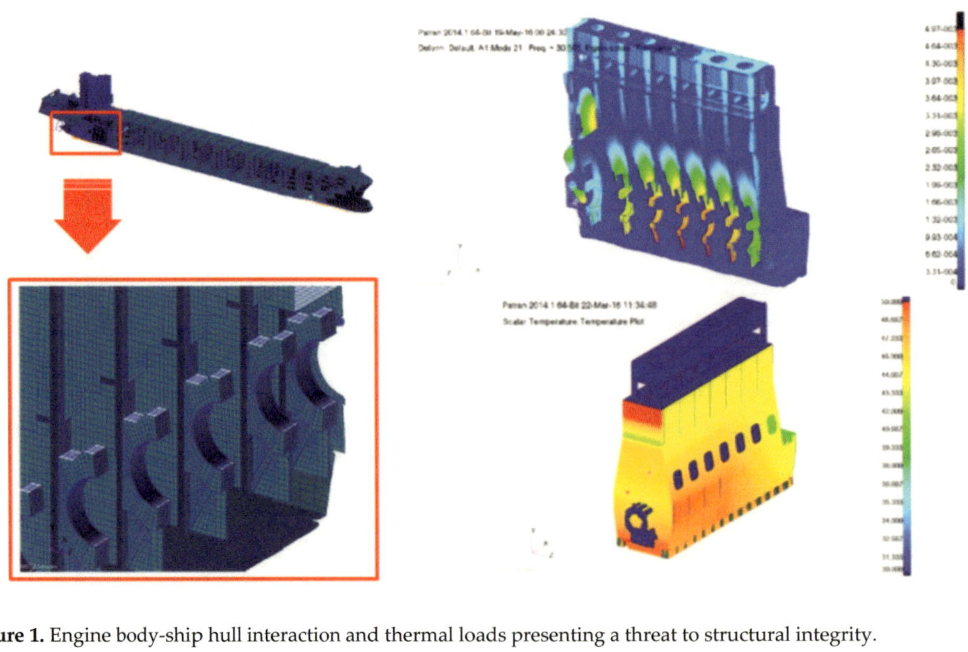

Figure 1. Engine body-ship hull interaction and thermal loads presenting a threat to structural integrity.

and direction change as well as thrust force direction change. The resulting effect is a dynamic load on collar coupling bolts in a longer operating time [10], which can result in fatigue failure. The fretting that occurs on adjacent connecting surfaces in these cases creates micronotches that develop into fatigue cracks with the direction of failure growth in planes angled from 35 to 60° which is not a characteristic of pure torsional fatigue failures. The analysis showed that the coupling bolts are subjected to an increasing bending moment which contributes to fatigue crack growth. The experimental research and numerical calculation done in this case study proved the hypothesis of variable bending stress in the coupling as the failure cause. Bolted connections of propeller blades and the shaft are often in a cathodic protection environment. Hydrogen inclusions in the material and variable stress conditions can cause crack nucleation and propagation, finally causing a failure [11]. Fractographic analysis, chemical analysis, microhardness tests, slow strain rate test, microstructure analysis, and finite element analysis were performed in this case.

Abnormal performance of the propeller by way of one non-performing malformed blade can generate a uniaxial force which fluctuates once per rotation in a consistent transverse direction across the shaft. The fluctuating force generates a couple which can cause fatigue failure of the propeller hub [12]. Uniaxial type of failure is characterized by a fatigue fracture with a single origination point that progresses across the shaft from the side where the force is being applied and results in the final overload failure occurring on the opposite side from the fluctuating force. Visual inspection, detail axis alignment measurements, microscopic metallurgical examination, hardness measurements, and ultrasonic scanning were used during the analysis.

2.3. Offshore structural failure

Offshore structures can be divided into three groups: fixed platforms (steel template and concrete gravity structures), compliant tower (compliant, guyed and articulated tower, and tension leg platform), and floating structures (floating production, storage and offloading systems).

The loads on offshore structures are gravity (self-weight, various equipment, fixed platform elements, and fluid loads), environmental (winds, waves, currents, and ice), exploitation loads, and seismic loads. Environmental loads play a major role in offshore structures design process.

In complex structures, such as offshore platforms, a fatigue failure of a single structural element may not result in a catastrophic failure of the entire structure, but it definitely changes the expected lifetime of the structure. The need for structural system failure probability estimation of typical marine structures in combination of fatigue and fracture arises. A proposed numerical and analytical method had been tested on real structures, like a Neka jack-up platform (Iran Khazar) [13], by applying various fatigue sequences that could lead to the collapse of the platform structure. This comparison has shown that the calculated system failure probability is higher for the case of combined fatigue and fracture scenarios than for only fatigue or fracture induced structure collapse which emphasizes the need for regular inspections of marine structures.

Offshore pipelines are usually damaged in the form of dents and gouges, which reduces its static and dynamic load bearing capacity as well as the fatigue life reduction in comparison to undamaged pipelines. The extent of the fatigue lifetime change depends on the type of the dent, and it can be analyzed and assessed analytically or numerically (FEM) [14]. Fatigue life analysis helps in the decision on the necessity of repairs and/or replacement of the damaged pipelines, i.e., planning of inspection and maintenance activities. Offshore pipelines segments are usually connected by welds which usually contain surface of embedded defects which exhibit large plastic strain characteristics if fracture occurs. In such cases, nonlinear elastic plastic fracture response should be modeled [15].

Subsea structures are subjected to significant external pressure loads which makes structural buckling a dominant failure mechanism. Ultra-deep water subsea separators are key equipment of subsea production in offshore petroleum industry. An experimental and numerical investigation on buckling and post-buckling of a 3000 m subsea separator has been done by Ge et al. [16]. The analysis has shown that the buckling behavior of deep sea structures can be assessed accurately applying numerical nonlinear global buckling analysis, proven by the comparison with experimental analysis results.

2.4. Marine equipment failure

This section deals with failures of marine equipment such as port or dock cranes, cables and ropes, pressure vessels-mounted onboard ships, and underwater pipelines.

Cranes can be subject to unexpected sudden events which can be divided into accidents and emergencies. Catastrophic failure of a dockside crane jib [17] occurred in the proximity of the standing tower, near the connection of the jib's three main tubes to the tower. Upon the visual inspection of the fracture surfaces, the presence of a large pre-existing crack was evident. The crack originated from a seam weld and propagated through one of the main pipes of the crane jib space frame. The failure occurred during maneuvering with no load attached. During the investigation crane material properties were obtained experimentally (tensile tests and Charpy impact tests) and the crane design was verified by FE analysis. Fatigue analysis was conducted, according to standards (FEM 1.001, Eurocode 3), for the welding joints and the pipes. Failure mode analysis was done from fracture mechanics and plastic collapse approaches. All of the analysis and investigations brought to the conclusion that the fatigue design of the jib structure was not done according to standards and that the final failure was determined by plastic collapse, after a long stable propagation period of a dominant crack which originated at the edge of a seam weld.

As for the pressure vessel failures, there are two main reasons for failures, i.e., pressure part failure (safety valves failures, corrosion, and low water level) or fuel/air explosions in the furnace (gas or liquid fuel leaks). Inadequate construction characteristics of high pressure tubes can cause failures. An investigation of a prematurely ruptured high-pressure oil tube has shown that inadequate pipe type (longitudinally welded instead of seamless) and material (design specified material replaced by a lower grade one) as well as inadequate installation procedures (not enough pipe clamps which allowed vibrations) resulted in vibration induced fatigue crack [18].

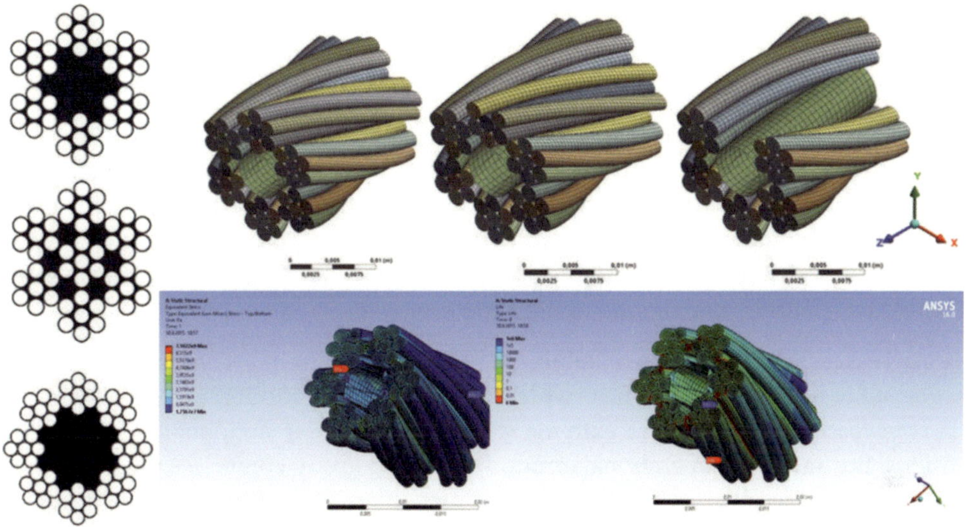

Figure 2. Numerical analysis of remaining fatigue life of a wire rope.

All equipment on marine structures is maintained and serviced continuously. In case of a malfunction, *in-situ* repairs are often performed. The quality of workmanship and material choice do have a great importance in such cases. Bending stresses in equipment elements that should be subjected only to tensile stress (ropes, wires, etc.) can cause failure of such elements. Numerical analysis of different wire rope cross section configurations is performed in order to determine remaining fatigue of operating wire ropes in dockside cranes [19], **Figure 2**.

Subsea umbilicals are composite cable and small diameter tubular bundles deployed on the seabed in conjunction with offshore installations for oil or gas exploitation. These tubes are loaded by alternating internal pressure and exposed to sea currents, i.e., dynamic loading [20]. Cracks in this type of equipment result in leaks and loss of load-carrying capacity. Umbilical tubes experience loss of circularity in shape (ovalization) and are subjected to re-rounding procedures by applying boost pressure prior to service which also translates in fatigue loading.

3. Failure causes and mechanisms

The strength of a structure represents a limit state of loading conditions above which the structure loses ability to achieve its specified required function. As long as the actual strength of the structure is kept higher than the actual loading demands, a given marine structure can be deemed safe. Otherwise, structural failures will occur.

Structural failure can be defined as loss of the load-carrying capacity of a component or member within a structure or of the structure itself (including global failure modes like capsizing, sinking, positioning system failures, etc.). The failure can result in catastrophic damage (i.e., complete loss of the structure itself) or partial structure damage when the structure can be repaired or recovered. Global failures can more often result in fatal casualties, while smaller and localized structural damage may result in pollution and recoverable structural damage.

Structural failure is initiated when the material in a structure is stressed to its strength limit, thus causing fracture or excessive deformations. The structural integrity of a marine structure depends on load conditions, the strength of the structure itself, manufacturing and materials quality level, severity of service conditions, design quality as well as various human elements that have effects during exploitation of the structure.

There are two distinctive groups of failure causes. The first group is comprised of unforeseeable external or environmental effects which exert additional loading on the structure resulting in over-load. Such effects are extreme weather (overloads), accidental loads (collisions, explosions, fire, etc.), and operational errors. The second group comprises the causes for failures that occur either during the design and construction phase (dimensioning errors, poor construction workmanship, material imperfections) or due to phenomena growing in time (fatigue), both resulting in reduced actual strength in respect to the design value. All of the listed causes can partially or completely be a result of human factor.

The process of fatigue failure itself is highly complex in nature and it is dependent on a large number of parameters. The factors are numerous and perhaps the most significant are mean stress (distribution), residual stresses, loading characteristics and sequence, structural dimensions, corrosion parameters, environmental temperature, design criteria fabrication methods and quality.

Failure mechanisms that usually occur in marine structures can be progressive (excessive yielding, buckling, excessive deformations) or sudden (brittle and fatigue fractures). Excessive yielding and brittle fractures occur when the load exceeds critical strength, while buckling and fatigue fractures depend on time and specific load conditions.

4. Failure analysis tools

The analysis methods can be grouped into methods that use nominal stresses (typical for standard codes) acting to a structure or part of a structure and then compare the stress amplitude to nominal S-N curves. This approach is appropriate for structures that are standardized, and therefore well backed up with statistical experimental data that can be used as initial assumptions for fatigue analysis. The alternative is the evaluation of local stresses influence to fatigue (notch stress factors, N-SIF).

Some authors [21] divide fatigue analysis methods in two groups: S-N approach based on fatigue tests and fracture mechanics approach. The first method is used for fatigue design purpose using simplified fatigue analysis, spectral fatigue analysis, or time domain fatigue analysis to determine fatigue loads. The second method is used for determination of acceptable flaw size, prediction of crack growth behavior, planning maintenance of the structure, and similar activities.

The latest trend in failure analysis development is the unification of analysis methods and procedures [22–24], in order to obtain a comprehensive procedure of structural failure analysis that would cover main failure modes and enable a safer and more efficient design, manufacture and maintenance processes.

4.1. Experimental tools

Nondestructive testing and examination (NDT and NDE), as well as structural health monitoring (SHM), of structures play a significant role in fracture analysis and control procedures. Any method used must not alter, change, or modify the failed condition, but must survey the failure in a nondestructive mode so as to not impact, change, or further degrade the failure zone. This kind of examination provides input values for fracture analysis, which yields results that define inspection and maintenance intervals for the structure and represent input values for life prediction estimates. Structures are inspected at the beginning of their service life in order to document initial flaws which determine the starting point of the structure fatigue life prediction. The most commonly used procedures for marine structures are optical microscopy, scanning electron microscopy (SEM), GDS, and acoustic emission (AE) testing.

Optical microscopy is a common and most widely used NDT analysis method which enables rapid location and identification of most external material defects. This technique is often used in conjunction with micro-sectioning to broaden the application. One of the main disadvantages is the narrow depth-of-field, especially at higher magnifications.

Scanning electron microscopy is an extension of optical microscopy in failure analysis. The use of electrons, instead of a light source, provides much higher magnification (up to 100,000×) and much better depth of field, unique imaging, and the opportunity to perform elemental analysis and phase identification. The examined item is placed in a vacuum enclosure and exposed with a finely focused electron beam. The main advantage of this method is minimal specimen preparation activity due to the fact that the thickness of the specimen does not pose any influence to the analysis, ultra-high resolution, and 3D resulting appearance of the test object. Various analyses of marine structures and equipment have been conducted using SEM [25–28], one of them being analysis of speed boat steering wheel fracture, **Figure 3**.

As it is well known, structural supporting members emit sounds prior to their collapse, i.e., failure. This fact has been the basis of the development of scientific methods of monitoring and analysis of these sounds with the goal to detect and locate faults in mechanically loaded structures and components. AE provides comprehensive information on the origin of a discontinuity (flaw) in a stressed component and also provides information about the development of flaws in structures under dynamic loading. Discontinuities in stressed components release energy which travels in the form of high-frequency stress waves. Ultrasonic sensors (20 kHz–1 MHz) receive these waves or oscillations and turn them into electronic signals which are in turn processed on a computer yielding data about the source location, intensity frequency spectrum, and other parameters that are of interest for the analysis. This method is passive, i.e., no active source of energy is applied in order to create observable effects as in other NDT methods (ultrasonic, radiography, etc.). Three sources of acoustic emissions are recognized, namely primary, secondary, and noise. The primary sources have the greatest

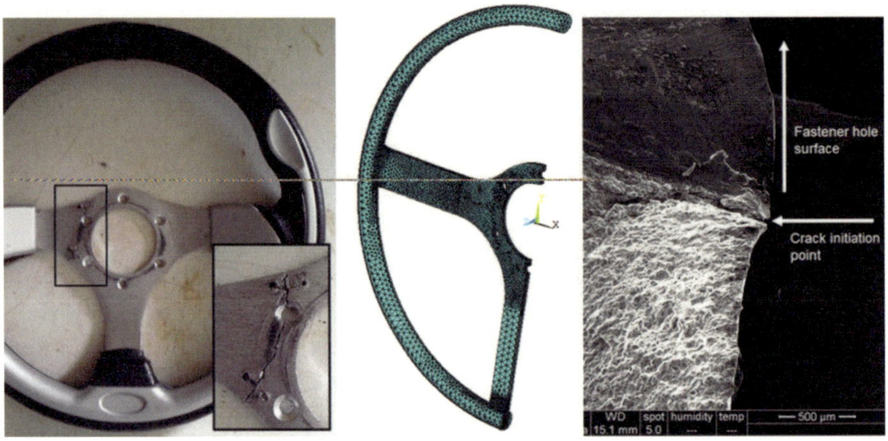

Figure 3. Experimental analysis of fractured speed boat steering wheel coupled with numerical analysis.

structural significance and originate in permanent defects in the material that manifest as local stresses, either on microstructural or macrostructural level. The amount of acoustic emission energy released, and the amplitude of the resulting wave, depends on the size and the speed of the source event. The main advantages of AE compared to other NDT methods are that AE can be used in all stages of testing. Additionally, it is less sensitive to changes in geometry, the scanning is remote and it gives real-time evaluation [29]. The disadvantages are the sensitivity to signal attenuation in the structure, less repeatability do to the uniqueness of emissions for a specific stress/loading conditions, and external noise influence on accuracy.

4.2. Analytical tools

Although various analytical models have been proposed by a number of authors, no comprehensive model exists. Analytical methods have been developed for prediction of progressive structural failures of marine structures [30]. The finite element modeling approach for prediction of the development of failures is accurate, but can be time consuming. Analytical procedures, based on spectral fatigue analysis, beam theory, fracture mechanics, and structural factors, can provide solutions in considerably less time when needed.

The goal is to define approaches for computing the fracture driving force in structural components that contain cracks. The most appropriate analytical methodology for a given situation depends on geometry, loading, and material properties. The decisive choice factor is the character of stress. If the structure behavior is predominantly elastic, linear elastic fracture mechanics can yield acceptable results. On the other hand, when significant yielding precedes fracture, elastic-plastic methods, such as referent stress approach (RSA) and failure assessment diagram (FAD), need to be used. Since a purely linear elastic fracture analysis can yield invalid and inaccurate results, the safest approach is to adopt an analysis that spans the entire range from linear elastic to fully plastic behavior. One of the methods that can be applied is the FAD approach.

The FAD approach has first been developed from the strip-yield model and it uses two parameters which are linearly dependent to the applied load. This method can be applied to analyze and model brittle fracture (from linear elastic to ductile overload), welded components fatigue behavior, or ductile tearing. The stress intensity factors are defined on the basis of the structure collapse stress and the geometry dependence of the strip-yield model is eliminated [31, 32]. The result is a curve that represents a set of points of predicted failure points, hence the name failure assessment diagram. The failure assessment diagram is basically an alternative method for graphically representing the fracture driving force.

Depending on the type of the equation used to model the effective stress intensity factors the FAD approach can be sub-divided into the strip-yield based FAD, J-based FAD, and approximated FAD. The J-based FAD includes the effects of hardening of the material, while the simplified approximations of the FAD curve are used to reduce the calculation times of the analysis. When stress-strain data are not available for the material of interest generic FAD expressions may be used [33], which assume that the FAD is independent of both geometry and material properties. The simplified curves proved adequate for most practical applications due to the fact that design stresses are usually below yield point. Fracture analysis in fully plastic regime requires an elastic-plastic J analysis.

Marine structures are subjected to dynamic load that are characterized by exactly unpredictable, stochastic changes of value (environmental factors). Most fracture mechanics analyses are deterministic, therefore a need to view fracture probabilistically for real world conditions arise. The probabilistic fracture analysis overlaps the probability distributions of driving force in the structure and toughness distribution in the structure to obtain a finite probability of failure. Probabilistic methods can take into account time-dependent crack growth and stress corrosion cracking by applying appropriate distribution laws. Most practical situations exhibit randomness and uncertainty of the analysis variables so numerical algorithms for probabilistic analysis may be needed to apply. The well-known Mote Carlo method has been proven to be suited to accompany FAD models in cases of uncertainties.

Recently, normative institutions have been involved in projects and research, together with industry, in order to establish probabilistic methods for planning in-service inspection for fatigue cracks in offshore structures. DNV-issued recommendations on how to use probabilistic methods for jacket structures, semisubmersibles, and floating production ships [34]. Basically, the goal of probabilistic method is to replace inspection planning based on engineering assessment of fatigue and failure consequences with mathematical models for the influence of exploitation, fatigue causes, and crack propagation characteristics on the lifetime of the structure to obtain a more reliable and secure assessment methodology independent of the engineers' level of expertise.

Li and Chow [35] have developed a fatigue damage model by formulating a set of damage coupled constitutive and evolution equations in order to write a computer software that could predict the behavior of offshore structures under dynamic load. The fatigue damage model is based on sea wave's characteristics statistics. The model also includes historical damage data.

Cui [36] has focused his research on the requirement for accurate fracture growth predictions that preceding fatigue strength assessment methods, mainly based on cumulative fatigue damage theory using stress-endurance curves (S-N), have not taken into account. The effects of initial defects and load sequence are included in the prediction model. A fatigue crack propagation theory has been proposed as technically feasible and adoptable method for fatigue life prediction using commercial FEA/FEM software packages for the calculation processes. The need for a database of the size and distribution of initial defects for marine structures is emphasized.

Li et al. [37] have developed an improved procedure for creation of standardized load-time history for marine structures based on a short-term load measurement. The need for load-time history arises from the dependency fatigue crack growth behavior to load sequence effect.

It is known that small variations in the initial (basic) assumptions for a fatigue analysis can have significant influence for the predicted crack growth time. As mentioned above, the S-N based calculations are sensitive to input parameters values and definitions [38]. As the occurrence of a crack is not strictly deterministic, probabilistic methods for the prediction of crack behavior and sizes, based on fatigue crack propagation theory, can resolve accuracy problems. Probabilistic methods require extensive database of standardized load-time histories for marine structures, based on extensive experimental research, which can be used in analysis procedures.

4.3. Numerical tools

The effective application of numerical methods in fracture mechanics and fatigue analysis begun with the development of computer science in the second half of the twentieth century. Various methods were used (finite difference method, collocation methods, and Fourier-transformations) but the finite element method (FEM) has been established as a standard due to its universality and efficiency. FEM enables complicated crack configuration analysis under complex loads and non-linear material behavior.

Recent years have brought a significant development and increase in accessibility of commercial computational software and hardware for finite element analysis applications, marine structures included. This enables more advanced and detailed fatigue and fracture analysis even for more complex large-scale structures. Furthermore, numerical tools can be used to complement or even substitute experimental analyses, as in the material selection stage in design process [39], **Figure 4**.

As the extent of scientific material published on this matter is very ample, here recently developed methods will be briefly described and referenced.

Extended FEM (X-FEM) is the most recent finite element method developed and is used mainly for fracture mechanics applications. Based on the finite element method and fracture mechanics theory, X-FEM can be applied to solve complicated discontinuity issues including fracture, interface, and damage problems with great potential for use in multi-scale computation and multi-phase coupling problems. The method has been introduced in 1999 [40] and since then further developed by various authors. The basic idea of the method is to reduce the re-meshing around the crack to a minimum. The improvements enabled the crack to be represented in the FE model independently from the mesh itself [40, 41]. The solution for the

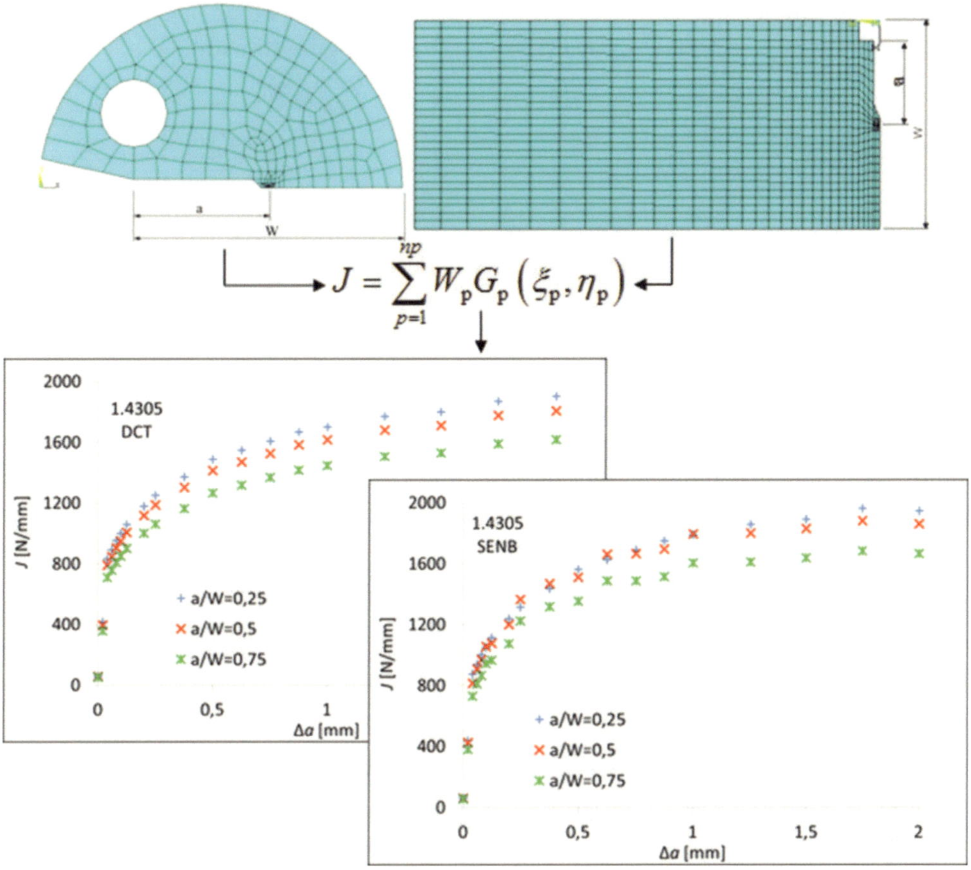

Figure 4. Numerical prediction of material fracture behavior using FE models of fracture mechanics standardized specimens in order to get dependence of J-integral to crack growth.

problem of modeling curved cracks was developed by forming higher order elements [42]. Improved XFEM methods are continuously being developed by various researchers as the method has been proven as very valuable.

Various computer software packages for fatigue crack growth analysis have been developed by NASA. FASTRAN is a life-prediction code based on the crack-closure concept and is used to predict crack length against cycles from a specified initial crack size to failure for many common crack configurations found in structural components. NASA FLAGRO v2 fatigue crack growth computer program developed as an aid in predicting the growth of pre-existing flaws and cracks in structural components using a two-dimensional model which predicts growth independently in two directions based on the calculation of stress intensity factors.

Recently, specific numerical automatic crack box technique (CBT) has been developed in order to enable to perform fine fracture mechanics calculations in various structures without global re-meshing [43]. The algorithm can be used for FEM calculations with ABAQUS code. The method represents an improvement as only the specific crack zone has to be re-meshed which results in simpler and time saving calculations. Also, the method allows the analysis of the influence of plastic material characteristics on the crack growth path.

5. Conclusions

This chapter provided an overview of common failures of marine structures taking into account failure mechanisms and tools used for failure analysis. As shown, the majority of employed failure analysis is comprised of visual, analytical, and mechanical inspection methods in the attempt to identify failure causes. The working conditions in which marine structures operate are often stochastic in nature and strongly dependent on weather conditions at sea as well on loading conditions of the structure. The complexity of failure analysis accentuates the need for numerical simulation of possible catastrophic scenarios during the entire lifetime span of the structure. If the marine structures coupled with the relevant data collected during maintenance procedures are numerically modeled than a tool for failure prediction can be developed. Therefore, complete analysis comprising analytical, experimental, and numerical research is desirable to obtain satisfying results.

Acknowledgements

The materials and data in this publication have been obtained through the support of the International Association of Maritime Universities (IAMU) and The Nippon Foundation in Japan.

Author details

Goran Vukelić* and Goran Vizentin

*Address all correspondence to: gvukelic@pfri.hr

Faculty of Maritime Studies Rijeka, University of Rijeka, Rijeka, Croatia

References

[1] Zhang W. Technical problem identification for the failures of the liberty ships. Challenges. 2016;7:20. DOI: 10.3390/challe7020020

[2] Class NK. Investigation Report on Structural Safety of Large Container Ships. Tokyo: Class NK; 2014.

[3] CLCSS. Final Report of Committee on Large Container Ship Safety (English Version) Tokyo: CLCSS; 2015. pp. 1-30

[4] Martins RF, Rodrigues H, Leal das Neves L, Pires da Silva P. Failure analysis of bilge keels and its design improvement. Engineering Failure Analysis. 2013;**27**:232-249. DOI: 10.1016/j.engfailanal.2012.06.002

[5] Zunkel A, Tiebe C, Schlischka J. "Stolt Rotterdam"—The sinking of an acid freighter. Engineering Failure Analysis. 2014;**43**:221-231. DOI: 10.1016/j.engfailanal.2014.03.002

[6] Grubišić V, Vulić N, Sönnichsen S. Structural durability validation of bearing girders in marine diesel engines. Engineering Failure Analysis. 2008;**15**:247-260. DOI: 10.1016/j.engfailanal.2007.01.014

[7] Murawski L. Thermal interaction between main engine body and ship hull. Ocean Engineering. 2018;**147**:107-120. DOI: 10.1016/j.oceaneng.2017.10.038

[8] Han HS, Lee KH, Park SH. Parametric study to identify the cause of high torsional vibration of the propulsion shaft in the ship. Engineering Failure Analysis. 2016;**59**:334-346. DOI: 10.1016/j.engfailanal.2015.10.018

[9] Arisoy CF, Başman G, Şeşen MK. Failure of a 17-4 PH stainless steel sailboat propeller shaft. Engineering Failure Analysis. 2003;**10**:711-717. DOI: 10.1016/S1350-6307(03)00041-4

[10] Dymarski C. Analysis of Ship Shaft Line Coupling Bolts Failure. Journal of Polish CIMAC. 2009;**4**(2):33-40

[11] Zhenqian Z, Zhiling T, Chun Y, Shuangping L. Failure analysis of vessel propeller bolts under fastening stress and cathode protection environment. Engineering Failure Analysis. 2015;**57**:129-136. DOI: 10.1016/j.engfailanal.2015.07.013

[12] Aurecon New Zealand Limited. Aratere Shaft Failure Investigation. Wellington: Aurecon New Zealand Limited; 2015

[13] Shabakhty N. System failure probability of offshore jack-up platforms in the combination of fatigue and fracture. Engineering Failure Analysis. 2011;**18**:223-243. DOI: 10.1016/j.engfailanal.2010.09.002

[14] Macdonald KA, Cosham A, Alexander CR, Hopkins P. Assessing mechanical damage in offshore pipelines—Two case studies. Engineering Failure Analysis. 2007;**14**:1667-1679. DOI: 10.1016/j.engfailanal.2006.11.074

[15] Zhang YM, Yi DK, Xiao ZM, Huang ZH. Engineering critical assessment for offshore pipelines with 3-D elliptical embedded cracks. Engineering Failure Analysis. 2015;**51**: 37-54. DOI: 10.1016/j.engfailanal.2015.02.018

[16] Ge J, Li W, Chen G, Li X, Ruan C, Zhang S. Experimental and numerical investigation on buckling and post-buckling of a 3000 m subsea separator. Engineering Failure Analysis. 2017;**74**:107-118. DOI: 10.1016/j.engfailanal.2017.01.001

[17] Frendo F. Analysis of the catastrophic failure of a dockside crane jib. Engineering Failure Analysis. 2013;**31**:394-411. DOI: 10.1016/j.engfailanal.2013.02.026

[18] Pardal JM, de Souza GC, Leão EC, da Silva MR, Tavares SSM. Fatigue cracking of high pressure oil tube. Case Studies in Engineering Failure Analysis. 2013;**1**:171-178. DOI: 10.1016/j.csefa.2013.07.001

[19] Vukelic G, Vizentin G. Damage-induced stresses and remaining service life predictions of wire ropes. Applied Sciences. 2017;**7**:107-113. DOI: 10.3390/app7010107

[20] Zerbst U, Stadie-Frohbös G, Plonski T, Jury J. The problem of adequate yield load solutions in the context of proof tests on a damaged subsea umbilical. Engineering Failure Analysis. 2009;**16**:1062-1073. DOI: 10.1016/j.engfailanal.2008.05.013

[21] Bai Y. Marine Structural Design. 1st ed. Amsterdam: Elsevier; 2003

[22] Gutiérrez-Solana F, Cicero S. FITNET FFS procedure: A unified European procedure for structural integrity assessment. Engineering Failure Analysis. 2009;**16**:559-577. DOI: 10.1016/j.engfailanal.2008.02.007

[23] Cui W, Wang F, Huang X. A unified fatigue life prediction method for marine structures. Marine Structures. 2011;**24**:153-181. DOI: 10.1016/j.marstruc.2011.02.007

[24] Choung J. Comparative studies of fracture models for marine structural steels. Ocean Engineering. 2009;**36**:1164-1174. DOI: 10.1016/j.oceaneng.2009.08.003

[25] Vukelic G. Failure study of a cracked speed boat steering wheel. Case Studies in Engineering Failure Analysis. 2015;**4**:76-82. DOI: 10.1016/j.csefa.2015.09.002

[26] Peng C, Zhu W, Liu Z, Wei X. Perforated mechanism of a water line outlet tee pipe for an oil well drilling rig. Case Studies in Engineering Failure Analysis. 2015;**4**:39-49. DOI: 10.1016/j.csefa.2015.07.002

[27] Ilman MN, Kusmono. Analysis of internal corrosion in subsea oil pipeline. Case Studies in Engineering Failure Analysis. 2014;**2**:1-8. DOI: 10.1016/j.csefa.2013.12.003

[28] Harris W, Birkitt K. Analysis of the failure of an offshore compressor crankshaft. Case Studies in Engineering Failure Analysis. 2016;**7**:50-55. DOI: 10.1016/j.csefa.2016.07.001

[29] Hellier CJ. Handbook of Nondestructive Evaluation. 2nd ed. New York: McGraw-Hill Education; 2013

[30] Bardetsky A, Lee A. Analytical prediction of progressive structural failure of a damaged ship for rapid response damage assessment. In: Proceedings of the ASME 2014 33rd International Conference on Ocean, Offshore and Arctic Engineering; 2016. pp. 1-9

[31] Dowling AR, Townley CHA. The effect of defects on structural failure: A two-criteria approach. International Journal of Pressure Vessels and Piping. 1975;**3**:77-107. DOI: 10.1016/0308-0161(75)90014-9

[32] Milne I, Ainsworth R, Dowling A, Stewart A. Assessment of the integrity of structures containing defects. International Journal of Pressure Vessels and Piping. 1988;**32**:3-104. DOI: 10.1016/0308-0161(88)90071-3

[33] Milne I, Ainsworth R, Dowling A, Stewart A. Background to and validation of CEGB report R/H/R6—Revision 3. International Journal of Pressure Vessels and Piping. 1988;**32**:105-196. DOI: 10.1016/0308-0161(88)90072-5

[34] Veritas DN. Probabilistic Methods for Planning of Inspection for Fatigue Cracks in Offshore Structures. 2015. p. 264

[35] Li DL, Chow CL. A damage mechanics approach to fatigue assessment in offshore structures. International Journal of Damage Mechanics. 1993;**2**:385-405. DOI: 10.1177/105678959300200405

[36] Cui W. A feasible study of fatigue life prediction for marine structures based on crack propagation analysis. Proceedings of the Institution of Mechanical Engineers, Part M. 2003;**217**:11-23. DOI: 10.1243/147509003321623112

[37] Li S, Cui W, Paik JK. An improved procedure for generating standardised load-time histories for marine structures. Proceedings of the Institution of Mechanical Engineers, Part M. 2016;**230**:281-296. DOI: 10.1177/1475090215569818

[38] Lotsber I, Sigurdsson G, Fjeldstad A, Torgeir M. Probabilistic Methods for Planning of Inspection for Fatigue Cracks in Offshore Structures. Marine Structures. 2016;**46**:167-192.

[39] Vukelic G, Brnic J. Numerical prediction of fracture behavior for austenitic and martensitic stainless steels. International Journal of Applied Mechanics. 2017;**9**:1750052-1-1750052-11. DOI: 10.1142/S1758825117500521

[40] Belytschko T, Black T. Elastic crack growth in finite elements with minimal remeshing. International Journal for Numerical Methods in Engineering. 1999;**45**:601-620. DOI: 10.1002/(SICI)1097-0207(19990620)45:5<601::AID-NME598>3.0.CO;2-S

[41] Moës N, Dolbow J, Belytschko T. A finite element method for crack growth without remeshing. International Journal for Numerical Methods in Engineering. 1999;**46**:131-150. DOI: 10.1002/(SICI)1097-0207(19990910)46:1<131::AID-NME726>3.0.CO;2-J

[42] Sonsino C. Fatigue testing under variable amplitude loading. International Journal of Fatigue. 2007;**29**:1080-1089. DOI: 10.1016/j.ijfatigue.2006.10.011

[43] Lebaillif D, Recho N. Brittle and ductile crack propagation using automatic finite element crack box technique. Engineering Fracture Mechanics. 2007;**74**:1810-1824. DOI: 10.1016/j.engfracmech.2006.08.029

X-Ray Techniques

Clementina Dilim Igwebike-Ossi

Additional information is available at the end of the chapter

http://dx.doi.org/10.5772/intechopen.72447

Abstract

This chapter reviewed existing X-ray techniques that can be used for the analysis of materials, inclusive of those used as engineering and structural components. These techniques are X-ray fluorescence (XRF) spectrometry, proton-induced X-ray emission (PIXE) spectrometry, and X-ray diffraction (XRD). These analytical techniques provide qualitative and quantitative information on the composition and structure of materials with precision. XRD gives information on the crystalline forms and amorphous content of materials, which could be quite useful in failure analysis if the type of failure brings about morphological changes in the material under investigation. PIXE and XRF provide information on the types of elements present in a sample material and their concentrations. PIXE is however preferable to XRF due to its higher sensitivity to trace elements and lower atomic number elements as well as its faster analysis. XRF and XRD are more commonly used than PIXE which is a powerful, high-tech method that is relatively new in the field of chemical research. In this chapter, the theory and principles of these analytical techniques are explained, and diagrams showing the components of spectrometers and diffractometers are provided with descriptions of how they function.

Keywords: X-rays, X-ray fluorescence (XRF), proton-induced X-ray emission (PIXE) spectrometry, X-ray diffraction (XRD), spectrometer, diffractometer

1. Introduction

Metallic and polymeric materials used as engineering and structural components can undergo mechanical failure due to a number of factors which include misuse, design errors or deficiencies, inadequate maintenance, overloading, and manufacturing defects [1]. The different types of failure a material can undergo include wear, fracture, fatigue, creep, stress, and corrosion [1], all of which affect the structural integrity and possibly the morphology of the material. Analyses of materials generally can be carried out using X-ray techniques, such as X-ray fluorescence (XRF) spectrometry, proton-induced X-ray emission (PIXE) spectrometry, and X-ray

INTECH
open science | open minds

© 2017 The Author(s). Licensee InTech. This chapter is distributed under the terms of the Creative Commons Attribution License (http://creativecommons.org/licenses/by/3.0), which permits unrestricted use, distribution, and reproduction in any medium, provided the original work is properly cited. (cc) BY

diffraction (XRD). A vast amount of information on elemental composition and concentration (XRF, PIXE), material morphology (XRD) can be obtained from X-ray techniques which make them very helpful in the analysis of materials generally.

X-rays were discovered in 1895 by the German physicist, Wilhelm Conrad Röntgen (1845–1923) who was later awarded the Nobel prize for physics in 1901 [2]. X-rays are invisible, highly penetrating electromagnetic radiation of much shorter wavelength (but higher frequency and energy) than visible light. The wavelength (λ) range for X-rays is from about 10^{-8} to 10^{-11} m [3, 4] and the corresponding frequency (ν) range is from about 10^{16} to 10^{19} s^{-1} [4]. The advantages of using X-rays in analysis are: (i) it is the cheapest and most convenient method. (ii) X-rays are not absorbed very much by air, so the specimen need not be in an evacuated chamber [3]. The disadvantage is that they do not interact very strongly with lighter elements, so this could impose a limitation on the elements detectable by X-ray techniques.

1.1. X-ray emission theory

When samples are bombarded (irradiated) with high-energy protons (or X-rays in the case of XRF and XRD), the interaction of the protons with the electrons of the atoms in the sample causes ejection of the electrons in the innermost shells in atoms of the specimen [4, 5]. This creates a hole (vacancy) in the inner shell, converting it to an ion thereby putting it in an unstable state. To restore the atoms to more stable states, i.e., their original configurations, the holes in the inner shells (or orbitals) are filled by electrons from outer shells. Such transitions from higher to lower energy levels are accompanied by energy emission in the form of X-ray photon, [4, 6] for instance an L-shell electron fills the hole in the K-shell. Since an L-shell electron has a higher energy than a K-shell electron, the surplus energy is emitted as X-rays. In a spectrum, this is seen as a line. The energy of the X-ray emitted when the vacancies are refilled depends on the difference in the energy of the inner shell with the initial hole and the energy of the electron that fills the hole. The emitted X-ray radiation is characteristic of the element from which they originate since each atom has its specific energy levels. The number of X-rays is proportional to the amount of the corresponding element within the sample. An energy dispersive detector is used to record and measure these X-rays and their intensities are then converted to elemental concentrations. An atom can have several lines due to electronic transitions and subsequent refilling of vacant holes by different electrons within the atom. This collection of lines is unique to that element and is like a fingerprint of the element.

1.2. X-ray fluorescence spectrometry

X-ray fluorescence (XRF) is a non-destructive analytical technique used for the identification of elements and determination of their concentrations in solid, powdered, and liquid samples [7, 8]. Elements present in samples are detectable by XRF up to 100% and at trace levels, usually below 1 part per million (ppm) [8]. The elements detectable by XRF range from Sodium to Uranium. XRF has inherent design limitations, which has reduced its sensitivity to lower atomic number elements making it unable to detect elements lighter than sodium. XRF is based on X-ray emission theory.

The wide application of XRF in industry and research is attributable to its ability to carry out accurate, reproducible analyses at very high speed. With modern, computer-controlled systems, operation is fully automatic and results are typically delivered within minutes or even seconds [8]. XRF analyzers are able to detect the elements present in a sample by measuring the secondary X-rays emitted from a sample irradiated with a primary X-ray source. Since a sample contains several elements, each of these elements produces a unique set of lines ("fingerprint") which is used in the identification of the element. This is why XRF spectrometry is a good technique for the analysis of the elemental composition of a material [9]. There are two types of XRF techniques: energy dispersive XRF (EDXRF), which has high accuracy and is sensitive for heavy metal analysis; and wavelength dispersive XRF (WDXRF), which is more suitable for the detection of light elements and rare earths [9].

1.2.1. How X-ray fluorescence works

X-ray fluorescence functions as follows [3, 7–9]:

1. The sample, which may be in the solid or liquid form, is bombarded with high-energy X-ray photons (primary X-rays) from an X-ray tube.
2. When an atom of an element in the sample is struck with an X-ray of sufficient energy (i.e., greater than the binding energy of the atom's K or L shell), an electron from one of the atom's innermost shells (K or L) is dislodged. This creates "holes" or vacancies in one or more of the orbitals, thereby, converting the atoms into ions which are unstable.
3. To restore stability to the atoms, the vacancies in the inner orbitals of lower energy levels are filled by electrons from outer orbitals, which are at higher energy levels. This transition from a higher to a lower energy orbital shell may be accompanied by an energy emission in the form of a secondary X-ray photon, a phenomenon known as "fluorescence." This is as a result of the release of excess energy by the higher orbital electron.
4. The energy (E) of the emitted fluorescence photons is determined by the difference in energies between the initial (higher) energy level (Ei) and final (lower) energy level (Ef) for the individual transitions. This energy difference (energy of the emitted photon) is related to the frequency (ν) of the photon by the mathematical expression:

$$\text{Energy (E) of emitted X}-\text{ray photon} = \text{Ei}-\text{Ef} = h\nu, \tag{1}$$

i.e., $E = h\nu$, but $c = \lambda\nu$, so $\nu = {}^{c}/_{\lambda}$

When the expression $\nu = {}^{c}/_{\lambda}$ is substituted into $E = h\nu$, we obtain the formula.

$$E = {}^{hc}/_{\lambda} \tag{2}$$

where h is Planck's constant = 6.62608×10^{-34} J s, c is the velocity of light = 2.9979×10^{8} m s^{-1}, and λ is the wavelength of the photon.

Thus, wavelengths are inversely proportional to the energies and are characteristic of each element.

5. The intensity of emission, i.e., number of photons is proportional to the concentration of the element responsible for the emission in a sample.

6. The measurement of the energy (E) of the emitted photons is the basis of X-ray fluorescence (XRF) analysis.

1.2.2. X-ray fluorescence spectrometer set-up

The set-up of the XRF spectrometer is shown in **Figure 1**. The major components are X-ray tube, diffractors (crystals), detectors, and counting electronics, which are described below.

1.2.2.1. X-ray tube

X rays can be produced in a highly evacuated glass bulb, called an X-ray tube, that contains two electrodes—an anode (positive electrode) and a cathode (negative electrode) [3]. The sealed X-ray tube is the primary radiation source and is powered by a high stability generator. The anode is usually made of platinum, tungsten, rhodium, or other heavy metals of high melting point [3, 8]. When a high voltage (about 40 KV) is applied between the electrodes, streams of electrons (cathode rays) are accelerated as they move from the cathode to the anode, producing X-rays as they strike the anode. The wavelength composition of the radiation from the X-ray tube depends upon the choice of anode material. For most applications, the optimal choice is a rhodium anode, although other options such as molybdenum, chromium, or gold may be preferable in certain circumstances [8].

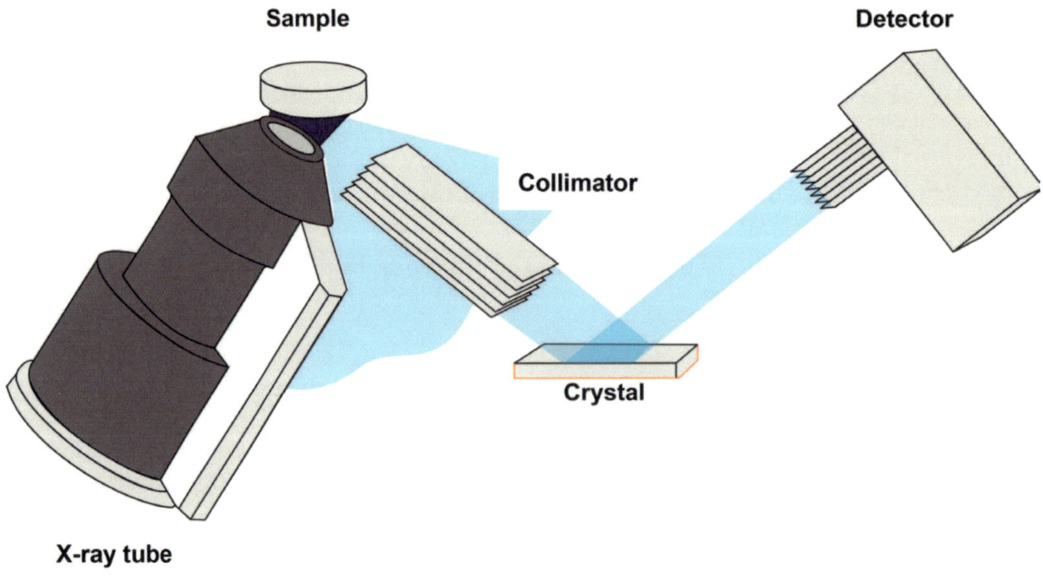

Figure 1. X-ray fluorescence spectrometer set-up (Source: PANalytical, the Analytical X-ray Company [8]).

1.2.2.2. Diffractors (crystals)

The separation of fluorescent X-ray peaks depends upon the relationship between wavelength and the d-spacing of the diffraction medium; consequently, a number of different crystals must be used to cover the full measurable range. Single crystals, such as germanium, lithum fluoride, and indium antimonide, are ideal diffractors for many elements [8]. More recently, synthetic multilayers with very small d-spacings have been introduced to provide enhanced sensitivity for the lighter elements [8].

1.2.2.3. Detectors

The detection of fluorescent radiation emitted is based on an ionization effect similar to that described under sample excitation. For the longer wavelengths produced by light elements, gas-filled proportional detectors are employed while short wavelengths (heavy elements) are measured with a scintillation detector. Both convert the photon energies into measurable voltage pulses [7, 8].

1.2.2.4. Counting electronics

Counting electronics record the number of pulses produced by the detectors and the energy levels corresponding to their amplitude [8]. Although data collection must continue for long enough to minimize statistical errors, measurement times as short as 2 s usually suffice for many elements. Longer times are required for the lightest elements, which produce relatively small numbers of low-energy fluorescent photons.

1.2.3. Interpretation of XRF spectra

An XRF spectrum is made up of XRF peaks with varying intensities. It is a graphical representation of X-ray intensity peaks as a function of energy peaks. The peak energy identifies the element, while the peak height/intensity is generally indicative of its concentration. An automated routine peak search or match identifies the elements present in unknown samples (qualitative) and their concentrations (quantitative).

1.2.4. Energy dispersive X-ray fluorescence (EDXRF)

EDXRF is the analytical technology commonly used in portable analyzers [9]. EDXRF is designed to analyze groups of elements simultaneously in order to rapidly determine those elements present in the sample and their relative concentrations. To understand how this method can be used, consider scrap metal. People in the business of recycling scrap metals need to positively identify numerous alloy grades, rapidly analyze their chemical composition at material transfer points, and guarantee the quality of their product to their customers [9]. This is important because metal alloys are designed for specific functions and are not interchangeable because small variations in composition can result in significantly different mechanical properties. However, hand-held XRF analyzers can easily separate these grades. Typical uses of EDXRF include the analysis of metals and alloys, petroleum oils and fuels, plastic, rubber and textiles, pharmaceutical products, foodstuffs, cosmetics, body care products, geological materials, cement, ceramics, etc. [9].

2. Particle induced X-ray emission (PIXE) spectrometry

2.1. Introduction

PIXE is an analytical method based on X-ray emission theory. Particle-induced X-ray emission or proton-induced X-ray emission (PIXE) spectrometry is a powerful, non-destructive analytical technique used to determine the elemental composition of a solid, liquid, thin film, and aerosol filter samples [5]. PIXE can detect all elements from sodium to uranium, giving a total of 72 elements (excluding Po, At, Fr, Ra, Ac, Pa, and the inert gases) detectable using this method [10]. PIXE technique relies on the analysis of the energy spectra of characteristic X-rays emitted by the de-excitation of the atoms in the sample bombarded with high-energy (1–3 MeV) protons with the aid of a suitable energy dispersive detector.

This technique was first proposed in 1970 by Sven Johansson at Lund University, Sweden, and developed over the next few years with his colleagues, Roland Akselsson and Thomas B. Johansson. PIXE is similar to other spectrometric technique used in elemental analysis as it is based on excitation of electrons in the atoms of the elements and electronic transitions that produce characteristic X-rays which by measurement of their intensities, the elements can be identified and their concentrations quantified [8]. The X-ray spectrum is initiated by energetic protons which excite the inner shell electrons in the target atoms. The expulsion of these inner shell electrons and re-filling of their vacant positions results in the emission of X-rays. The energies of the emitted X-rays are unique characteristic of the elements from which they originate, and the number of X-rays emitted is proportional to the mass/concentration of that corresponding element in the sample being analyzed [3, 5, 7, 8]. The generation of X-rays in a sample is very strongly influenced by the bombarding proton.

The use of proton beams for excitation has several advantages over other X-ray techniques. These are [5]:

- **i.** Higher sensitivity to trace elements.
- **ii.** Faster analysis due to higher rate of data accumulation across the entire spectrum.
- **iii.** Better sensitivity, especially for the lower atomic number elements.

The high sensitivity of PIXE in trace elements determination is due to a lower Bremstrahlung background, which is as a result of the deceleration of dislodged electrons when compared to electron excitation and the lack of a background continuum when compared with XRF analysis. The increasing need for elemental analysis of very small samples (0.1–1 mg) as in aerosol filters has made PIXE gain wide acceptance as a valuable analytical tool [5]. Samples (or materials) whose elemental make-up can be determined using the PIXE technique include oils and fuels, plastics, rubbers, textiles, pharmaceutical products, foodstuffs, cosmetics, fertilizers, minerals, ores, coals, rocks and sediments, cements, ceramics, polymers, inks, resins, papers, soils, ash, leaves, films, tissues, forensics, catalysts, etc. [10]. The most extensive use of the PIXE method, however, is in the elemental analyses of atmospheric aerosol samples, dust and fly ash samples, different biological materials, and archeological and artistic artifacts. The need for small accelerators in nuclear physics research laboratories four decades ago, has largely confined PIXE to such laboratories; it is yet to be widely used for analytical purposes.

2.2. Development of PIXE analytical technique

This new analytical method, which became known under the acronym PIXE, was tested and applied in many nuclear physics laboratories during the 1970s [5]. The development of PIXE has been quite rapid. There were several reasons for its rapid development; first, the growing global interest in environmental protection issues created a need for efficient methods of elemental analysis for air pollution studies and the determination of toxic elements in the environment and in humans. Because, PIXE is well suited for the determination of trace elements in a matrix of light elements, it is ideal for studies of this kind. Second, as a consequence of the early days of fundamental nuclear physics research, small accelerators became available in many nuclear physics laboratories, where they were the standard equipment. However, interest soon shifted to higher energies and the small machines became obsolete, as far as nuclear physics was concerned. An alternative to scrapping them was to use them for research in applied science, and PIXE was one of the most popular options. A contributory factor was that not only the accelerator, but also auxiliary equipment such as detectors, electronics, and computer facilities were available in most laboratories making it possible for feasibility tests to be carried out in many laboratories. PIXE is capable of detecting elemental concentrations down to parts per million; however, the technology is still relatively new and untested in wider avenues of chemical research.

2.3. Ion beam analysis (IBA) principles

PIXE is one out of the four ion beam analysis (IBA) methods. IBA consists of whole methods of studying materials based on the interaction at both atomic and the nuclear level, between accelerated charged particles (ions) and the bombarded material (sample) [11]. When a charged particle moving at high speed strikes a material, a number of events can take place. The ion can interact with the electrons and nuclei of the material atoms, slows down and possibly deviates from its initial trajectory. This can lead to the emission of particles and/or radiations (X- and γ-rays), whose energy is characteristic of the elements that constitute the sample.

2.3.1. Ion beam analysis methods

The spectrometric analysis of the various secondary emissions leads to the various IBA techniques [5, 11, 12]:

1. **PIXE (particle-induced X-ray emission)** is based on atomic fluorescence and the analysis is performed with characteristic X-rays. PIXE is well adapted for the analysis of trace elements ranging from Na to U.

2. **PIGE (particle-induced Gamma-ray emission)** is based on nuclear reaction and the analysis is performed with characteristic Gamma-ray. PIGE is particularly useful for analyzing light elements such as F and lighter elements, which are inaccessible by PIXE

3. **NRA (nuclear reaction analysis)** is based on nuclear reaction and the analysis is performed with charged particles. NRA has demonstrated its usefulness in the study of the oxidation and deposition of hydrocarbon residue on metallic surfaces.

4. **RBS (Rutherford backscattering analysis)** is based on nuclear scattering and the analysis is performed by charged particles. RBS has proved its efficacy in identifying and localizing thin layers.

2.4. PIXE analysis

2.4.1. Basic principles

As a charged particle (proton) moves through a material, it loses energy primarily by exciting electrons in the atoms that it passes by. Electrons in the inner shells of the atom (predominantly the K and L shells) are given enough energy to cause them to be ejected, resulting in an unstable atom (ion). Electrons from higher shells in the atom then "drop down" to fill the vacancies and in so doing, give off excess energy in the form of X-rays. The energies of these X-rays are characteristic of the element and therefore can be used to identify elemental composition. Also, by measuring intensities of characteristic X-ray lines, one can determine concentrations of almost all elements in the sample down to approximately 1 ppm (part per million).

2.4.2. Sample preparation

No special sample preparation is required in PIXE analysis as is the case in spectroscopic techniques (UV, IR). This minimizes the potential for error resulting from sample preparation. Most samples are usually analyzed in their original states, e.g., aerosol filter, archeological samples, soil, ash, and biological samples. However, it is very important that the area/volume of the sample irradiated by the beam (usually a circular area with the diameter of 1–10 mm) is representative of the whole sample. PIXE technique probes only the top 10–50 μm of the sample (depending on the material, energy of the incident beam, and most importantly, on the energy of characteristic X-rays), therefore if the sample is not homogeneous, as is the case with some pottery and geological samples, it is advisable to grind the sample to a fine powder (with particle size less than 1–2 μm), thoroughly mix it with 20% analytical grade carbon powder and press into pellets. Samples for PIXE analysis may be in the form of solids, liquids, aerosol filters, and thin membranes. The samples which usually come in different forms are handled in different ways, the details of which are described below [10].

Solid materials such as plastics, papers, or metals are analyzed in their "as received" condition, while materials in powdered form such as fly ash, activated carbon, catalysts, and corrosion products are first ground to reduce the particle size to about 200 mesh or lower and pressed into pellets before analysis.

Liquid samples such as oils, process waters, and solutions are analyzed using a plastic cup of either 8 or 3 ml in capacity with a 0.3 mil Kapton front surface window and can be analyzed as received by this method without modification. However, some liquids that are highly caustic or highly acidic may require pre-dilution or neutralization before analysis.

Aerosol filters and thin-film membrane samples are prepared on a clean bench and environment then immediately transferred to the target chamber for analysis in order to eliminate chances of sample contamination. Sample preparation is done by simply placing the filters or membranes as they are received, into snap-together plastic holders, which are then placed in the sample carousel. Since no permanent mounting is used, samples may be returned intact to clients upon request for archiving or for further analysis.

2.5. CERD ion beam analysis (IBA) set-up (accelerator room)

The Centre for Energy Research and Development (CERD), Obafemi Awolowo University, Ile-Ife, Nigeria, acquired a 1.7 MeV Tandem Pelletron accelerator a few years ago. It is the first ion beam facility in Nigeria, and the only one in the West African sub-region [11]. This facility has provided the opportunity for CERD to apply ion beam analysis techniques in the accelerator laboratory, particularly, particle-induced X-ray emission (PIXE), Rutherford backscattering spectroscopy (RBS), particle-induced Gamma-ray emission (PIGE), and elastic recoil detection analysis (ERDA). However, the focus of this chapter is only on PIXE since it is based on the theory of X-ray emission, which makes it an X-ray technique that can be used in the analysis of materials.

The ion beam analysis (IBA) setup, generally known as the accelerator room, is presented in **Figure 2** The major components of the setup are the accelerator, end station, and detector [5, 11, 12].

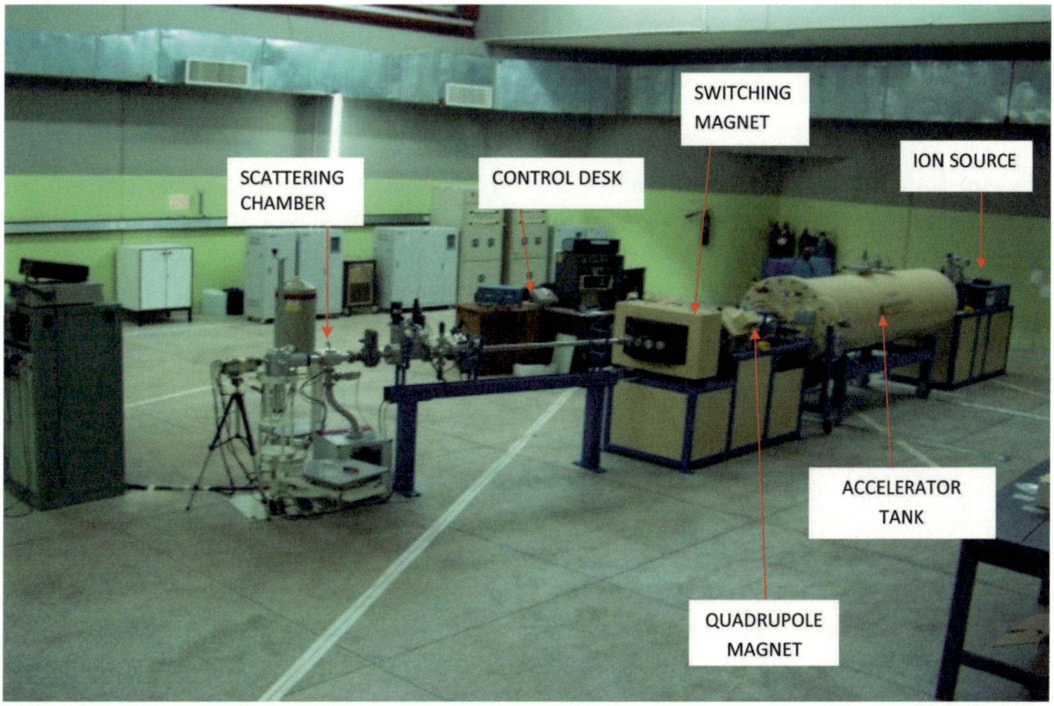

Figure 2. General view of ion beam analysis (IBA) setup at the Centre for Energy Research and Development (CERD), Obafemi Awolowo University, Ile-Ife, Nigeria (Source: [11, 13]).

2.5.1. *The accelerator*

The IBA facility is centered on a Tandem Pelletron Accelerator, Model 5SDH, built by the National Electrostatics Corporation (NEC), USA. It is equipped with an RF charge exchange ion source (Alphatross) to provide both proton and helium ions [12]. Positive ion beam is extracted from a plasma in the RF source and accelerated at 4.6 KeV for protons (6 KeV for alphas) into the charge exchange cell, where a portion (1–2%) is converted to negative ions by means of rubidium vapor. These negative ions are extracted and then accelerated to the desired energy by the Tandem Pelletron accelerator, which has a maximum terminal voltage of 1.7 MV [11]. At the terminal, in the center of the accelerator, a nitrogen stripper gas converts the negative ions to positive ions and they undergo a second-stage of acceleration. Thus the accelerator can deliver a proton beam of 0.6–3.4 MeV or an alpha beam of up to 5.1 MeV. The accelerator tank is filled with SF_6 insulating gas at a pressure of 80 psig. Two ultrahigh vacuum turbo molecular pumps rated at 3001/s, one at the low energy (LE) end and the other at the high energy (HE) end of the accelerator maintain the ultrahigh vacuum which could reach 2×10^8 Torr inside the accelerator tube and $\sim 10^{-7}$ Torr in the beam line extension [5, 11]. Control of the accelerator is facilitated by a computerized control panel with digital and analog displays. The accelerator has provision for five beam lines, but only PIXE is in use. The beam line in use (+15°) is equipped with a multi-purpose end-station for broad beam IBA analysis.

2.5.2. *The end station*

The general view of the end-station in CERD is shown in **Figure 3**. The end-station was designed and built by the Materials Research Group (MRG) at iThemba Labs, Sommerset West, South Africa [11].

The end-station consists of an Aluminum chamber of about 150 cm diameter and 180 cm height. At 90 cm height, the chamber has four ports and a window. Port 1 at 165° is for the RBS detector, Port 2 at 135° is for the PIXE detector, Port 3 at 30° is for the ERDA detector, the window at 0° is for observing the beam position and size, while port 4 at 225° is for PIGE [5, 11]. The chamber has a sample ladder that can accommodate eleven 13 mm samples. The chamber has a sample ladder that can accommodate 11 samples. The end-station has a turbo pump and a variable tantalum beam collimator (1, 2, 4, and 8 mm diameter) to regulate beam size and an isolation value.

2.5.3. *The detector*

The PIXE detector is a canberra Si (Li) detector (model ESLX30-150) with 30 mm^2 active area, a 25-mm thick Be window, and 150 eV FWHM energy resolution at 5.9 KeV [11]. There is a wheel in front of the PIXE detector to hold up eight different absorbers (four are installed) to cut off low energy peaks if necessary and/or reduce the count rates of the low Z elements. Canberra Genie 2000 (3.1) software is used for the simultaneous acquisition of the PIXE and RBS data. GUPIXWIN is the computer code used for PIXE data analysis.

http://dx.doi.org/10.5772/intechopen.72447

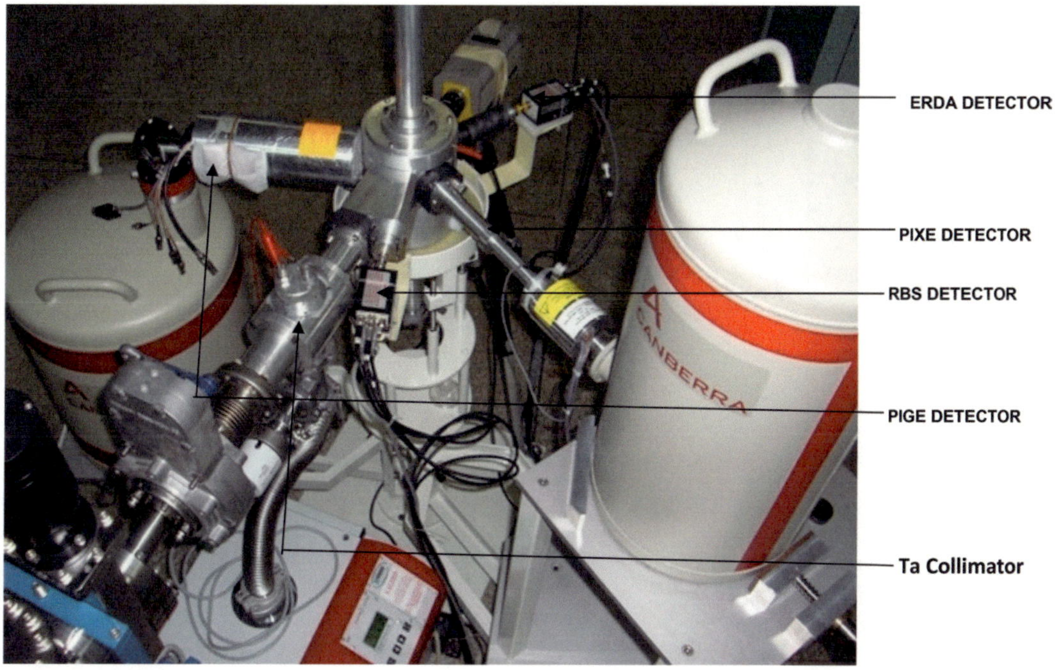

Figure 3. General view of the IBA End Station at CERD, OAU, Ife, Nigeria showing four ports and a window for RBS, PIXE, ERDA, and PIGE detectors (Source: [13]).

2.6. PIXE spectral results

PIXE spectra are generated in parts per million due to the sensitivity of the technique. The PIXE spectrum of rice husk ash heated at a combustion temperature of 600°C and duration of 5 h is presented in **Figure 4** as an example [13]. The corresponding concentrations values (ppm) of the elements in **Figure 4**, determined by computer software, are presented in **Table 1**.

A typical PIXE spectrum is composed of [11]:

1. Characteristic X-ray lines
2. Background
3. Spectrum artifacts

The X-ray energy spectrum consists of a continuous background together with the characteristic X-ray lines of the atoms present in the specimen. The X-rays are detected by means of a Si (Li) detector and the pulses from the detector are amplified and finally registered in a pulse height analyzer. Since a PIXE spectrum is usually quite complicated with many peaks, some of them overlapping, a computer is used for its deconvolution [5]. The number of pulses in each peak, which is a measure of the concentration of the corresponding element in the specimen, is calculated.

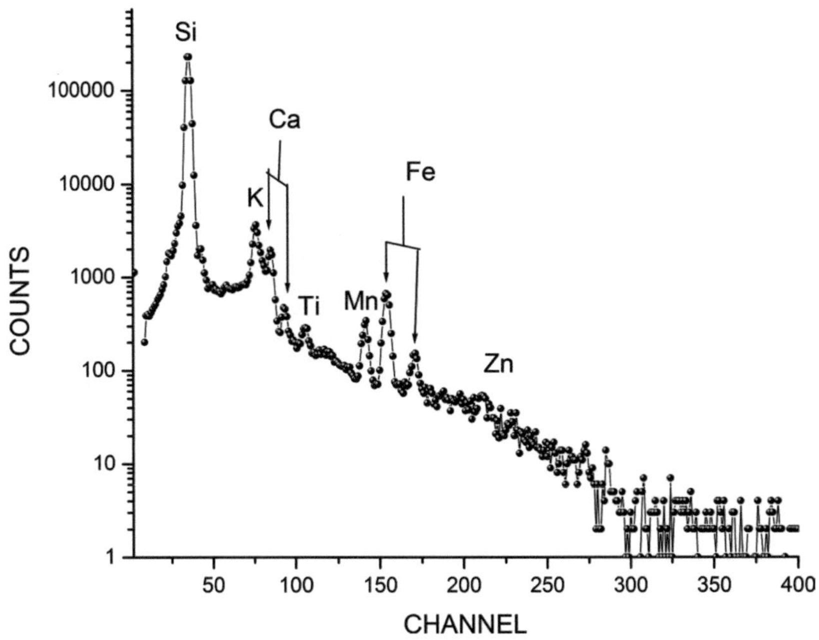

Figure 4. PIXE spectrum of rice husk ash obtained by heating rice husks at 600°C for 5 h. (Source: [14]).

Element (symbol)	Concentration (ppm)	Statistical error (ppm)
Mg	3344.4	245.48
Al	258.9	22.39
Si	358368.2	430.04
P	7995.4	410.96
S	201.0	93.36
Cl	279.9	56.85
K	2008.3	40.97
Ca	909.3	18.46
Ti	76.5	8.34
Mn	222.4	10.52
Fe	696.2	16.78
Zn	9.6	4.10

Table 1. PIXE concentration values (ppm) of the elements in rice husk ash obtained by heating rice husks at 600°C for 5 h [14].

http://dx.doi.org/10.5772/intechopen.72447

2.6.1. *PIXAN-PIXE analysis software*

PIXAN, the software used for the analysis of PIXE, covers the following areas [11]:

i. Spectrum analysis—Determination of peak areas.

ii. Estimation of elemental concentrations from peak areas; thin target, thick target.

iii. System calibration against standards.

PIXAN consists of five executable programs.

2.6.2. *Calculation of final concentrations*

For thin samples, e.g., filters: Conc (μg/cm²) = $\frac{PEAK\ AREA}{THIN}$

For thick samples, e.g., solid: Conc (mg/kg) = $\frac{PEAK\ AREA}{YIELD}$

Note: The spectrum of the elemental composition of rice husk ash shows a preponderance of silicon (present in the form of silica) in the ash as is evident from the peak height. Other elements present in small quantities are K, Ca, Mn, Fe, while Ti and Zn are present in minute quantities. With the aid of PIXAN software, the calculations of the concentrations of the elements shown in the spectrum were translated into figures in parts per million as shown in **Table 1**.

2.6.3. *Conversion of concentration values of elements from parts per million (ppm) to percentage*

It is often necessary, for practical purposes, to convert the PIXE concentration values from ppm to % concentration. This calculation is done as follows [13–16]:

$$1\ \text{ppm} = \frac{1}{1,000,000} \tag{3}$$

$$\text{In}\ \% = \frac{1}{1,000,000} \times 100 = \frac{1}{10,000} \tag{4}$$

$$\text{To convert X ppm to X}\ \% = \frac{Xppm}{10,000} \tag{5}$$

Example: If an element has a concentration value of 1354 ppm its concentration in % is 0.1354%.

2.6.4. *Conversion of concentration values of elements (in ppm) to their oxides in percentage*

PIXE analysis gives the concentration of the elements in parts per million (ppm). However, if the samples are in the form of ash, it means the elements present are in the form of oxides. To convert the concentration of the elements expressed in ppm to their oxides (also in ppm), the former is divided by a conversion factor, which is obtained from a ratio of the element to its oxide as follows [13–15]:

$$\text{Oxide of element (in ppm)} = \frac{concentration\ of\ element\ in\ ppm}{Conversion\ factor} \tag{6}$$

However to convert concentration from ppm to percentage (%), the value obtained above is divided by 10,000. For example, to convert Mg with concentration value of 4164.5 ppm to MgO (in %)

$$\begin{aligned} \text{MgO Concentration } (\%) &= \frac{4164.5\ ppm}{0.6031} \times \frac{1}{10000} \quad \text{where 0.6031 is the conversion factor} \\ &= \frac{6905.16}{10000} = 0.6905\,\% \text{ of MgO.} \end{aligned} \tag{7}$$

The conversion factor is obtained from the ratios of the atomic weights of the pure elements to their oxides. Thus, the ratio of atomic weight of Mg to that of MgO = $\frac{24.31}{40.31}$, which gives the value of 0.6031.

3. X-ray diffraction technique

3.1. X-ray diffraction

The most commonly used device for X-ray generation is an X-ray tube, which consists of a cathode that emits electrons and an anode (the target). When electrons from the cathode are accelerated by a high voltage and bombard a metal target, usually a heavy metal such as copper or molybdenum, X-rays are generated. When X-rays irradiate a material sample, in cylindrical or pellet form, scattering occurs. The nature of scattering of the X-rays by the sample is determined by the morphology (degree of crystallinity and amorphousness) of the sample and could be *coherent* or *incoherent*. In coherent scattering, also known as *X-ray diffraction* there is no change in wavelength or phase between the incident and scattered rays [17]. Crystalline samples cause coherent scattering, while amorphous and semi-crystalline samples bring about incoherent scattering also known as diffuse scattering, in which there is change in both wavelength and phase. Every crystalline substance scatters X-rays in its own unique diffraction pattern due to differences in planar spacing of the crystals. The diffraction pattern of crystalline materials is composed of a series of concentric cones arising from scattering by the crystal planes. As the degree of crystallinity increases, the rings become more sharply defined. Consequently, for predominantly amorphous materials, the X-ray diffraction pattern is diffuse and a halo (characterized by a dark-shaded portion in the center) is observable. A great deal of information on the morphology and structure of a material is obtainable from the visual inspection and mathematical interpretation of the pattern and intensity of the scattered radiation, such as degree of crystallinity, dimensions of crystalline domains, bond distances and angles, and type of conformations in the crystalline regions. In the case of mechanical failure of a material, the morphology may involve a transformation from a high degree of crystallinity to a lower degree or even amorphousness which will be evident in the X-ray diffraction patterns of the material before and after failure.

3.2. X-ray diffraction as an analytical technique

X-ray diffraction (XRD) is a versatile, non-destructive analytical technique for the identification and quantitative determination of the various crystalline forms known as "phases" of compounds present in powdered and solid samples [8]. X-ray diffraction is based on constructive interference of monochromatic X-rays and a crystalline sample. Identification is achieved by comparing the X-ray diffraction pattern or "diffractogram" obtained from an unknown sample with an internationally recognized database containing reference patterns for more than 70,000 phases [8]. The most commonly used XRD methods involve the use of single crystals, but *powder diffraction* techniques are also used, especially for investigating solids with infinite lattice structures [18]. X-rays are chosen for these techniques because they have wavelengths of the same order of magnitude as typical inter-atomic distances in crystalline solids [19]. Consequently, diffraction is observed when X-rays interact with an array of atoms in a solid. The process involves generation of X-rays by a cathode ray tube, filtration to produce monochromatic radiation and collimation to concentrate the rays before it is directed towards the sample [3, 4]. The interaction of the incident X-rays with the sample produces constructive interference when conditions satisfy Bragg's law. Modern computer-controlled diffractometer systems use automatic routines to measure, record, and interpret the unique diffractograms produced by individual constituents in even highly complex mixtures.

The result of an XRD measurement is a *diffractogram* showing: (i) phases present (peak positions), (ii) phase concentrations (peak heights), (iii) amorphous content (background hump), and (iv) crystallite size/strain (peak widths) [8]. The widths of the peaks in a particular phase pattern provide an indication of the average crystallite size. Large crystallites give rise to sharp peaks, while the peak width increases as crystallite size reduces. XRD can also be used to measure texture of sample, stress in sample, and for analysis of thin films.

A crystal lattice has a regular structure with three-dimensional distribution of atoms in space, e.g., cubic, rhombic. The atoms are arranged in such a way that they form a series of parallel planes separated from one another by a distance d, which varies according to the nature of the material. [8]. For any crystal, planes exist in a number of different orientations, each with its own specific d-spacing. Defects in the lattice structures of crystalline solids have an important and sometimes dominating influence on the mechanical, electrical, and optical properties of solid materials [19]. XRD technique can be used to obtain information on the defects and imperfections in the crystal lattice of solids which could be natural or as a result of failure.

XRD is useful in analyzing a wide range of materials, from powders and thin films to nanomaterials and solid objects. In powders, chemical phases are identified qualitatively as well as quantitatively. High-resolution X-ray diffraction reveals the layer parameters, such as composition, thickness, roughness, and density in semiconductor thin films.

3.3. Diffraction and Bragg's Law

Diffraction is a wave phenomenon in which there is apparent bending and spreading of waves when they meet an obstruction. Diffraction occurs with electromagnetic waves, such as light and radio waves as well as sound and water waves. Light diffraction is caused by light bending around the edge of an object. To understand diffraction, we have to consider what happens when a wave interacts with a single particle. The particle scatters the incident beam uniformly in all directions. If the beam is incident on a solid crystalline material, the scattered beams may add together in a few directions and reinforce each other to give diffracted beams. By studying the diffraction pattern of a beam of radiation incident on the crystal, the structure of the crystal can be determined. As light is diffracted by a grating, so does radiation beam take place only in a certain specific direction. Measurements of the directions of the diffraction and the corresponding intensities provide information on the crystal structure responsible for diffraction.

3.3.1. Bragg's Law equation

Bragg's Law states that when a monochromatic and coherent (in-phase) beam of X-rays is incident on a crystal surface at an angle θ, scattering occurs; constructive interference of the scattered rays also occurs at angle, θ to the planes if the path length (interplanar) difference, d, is equal to a whole number, n, of wavelengths. The angles of the crystal and detector can be varied so that a particular wavelength can be measured. The efficiency of the scattering depends on the number and distribution of the electrons at the lattice sites, which is determined by the structure of the molecules that occupy the lattice site. By varying the angle θ, the Bragg's Law conditions are satisfied by different d-spacings in polycrystalline materials.

Thus, the diffraction process occurs when the Bragg's Law condition is satisfied [18]. It is mathematically expressed as:

$$n\lambda = 2d\sin\theta \tag{8}$$

where λ is the wavelength of incident X-rays, d is the interplanar spacing, θ is the X-ray angle of incidence, and n is an integer.

This means that the two waves, originally in phase, have to remain in phase as they are scattered. This relationship between the wavelength, λ of incident X-ray radiation and the lattice spacings, d of the crystal is *Bragg's law* and is the basis for the technique of X-ray diffraction [4] (**Figure 5**). The interplanar spacing of a crystal lattice determines the angles at which strong X-ray diffractions occur. These interplanar spacings (also called lattice spacings) are an inherent characteristic of the crystal, for they are determined by the size and arrangement of its atoms. Each crystalline compound has its set of interplanar spacings and thus its own characteristic set of X-ray diffraction angles that like a fingerprint can be used to identify the substance [19] as illustrated in **Figure 5**.

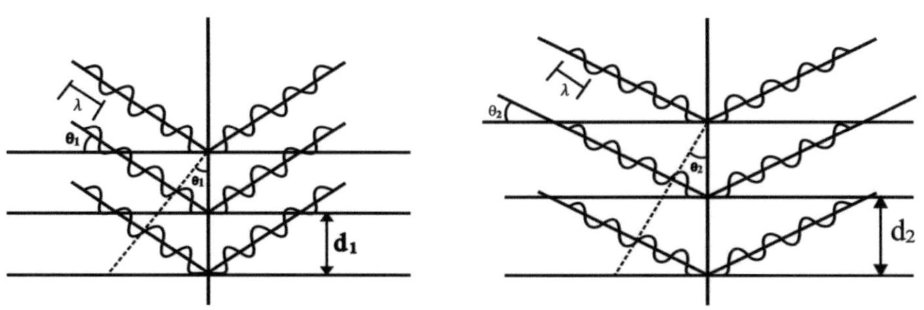

Figure 5. Diffraction: Bragg's Law [8].

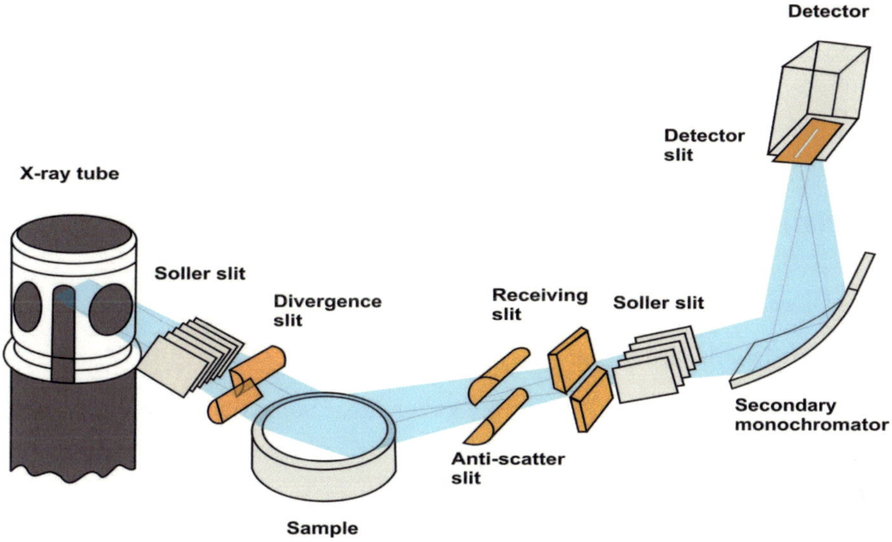

Figure 6. X-ray diffractometer set-up (Source: PANalytical, the analytical X-ray company [8]).

3.4. X-ray diffractometer set-up

An X-ray diffractometer typically consists of of an X-ray source, a mounting for the crystal, turntables to allow variation in the angles of the incident X-ray beam and crystal face, a slit arrangement, a monochromator and an X-ray detector [8, 18–20]. The setup of an X-ray diffractometer is shown in **Figure 6**.

Stresses and preferred orientation can be determined in a wide range of solid objects and engineered components. Many researchers in industrial as well as scientific laboratories rely on X-ray diffraction as a tool to develop new materials or to improve speed and efficiency of production processes. Fully automated X-ray diffraction analysis in mining and building materials production sites results in more cost-effective solutions for production control. Innovations in XRD closely follow research on new materials such as in semiconductor technologies and pharmaceutical investigations. XRD provides answers to many analytical questions related to the structure of material samples which makes it a useful tool in the analysis of failure in materials.

4. Conclusion

This review has shown that a great deal of information is obtainable with precision from the use of X-ray techniques in the analysis of materials. Analysis by XRD technique provides information for the identification and quantitative determination of the various crystalline forms of a material so it is useful in detecting the morphological changes that have occurred in a material after failure, if the crystal lattice structure has been affected. XRF and PIXE are both useful in the determination of the elemental components of a material and their concentrations.

However, PIXE, a powerful, high-tech analytical tool is preferable to XRF, due to its higher sensitivity to trace elements and lower atomic number elements as well as its faster analysis. However, if on-the-spot analysis is required as is sometimes the case in field work, EDXRF and WDXRF spectrometers would be preferred due to the advantage of portability.

Author details

Clementina Dilim Igwebike-Ossi

Address all correspondence to: clemdossi@yahoo.com

Department of Industrial Chemistry, Faculty of Science, Ebonyi State University, Abakaliki, Nigeria

References

[1] Maleque MA, Salit MS. Mechanical failure of materials. In: Materials Selection and Design. Singapore: Springer Briefs in Materials; 2013. p. 17

[2] Dunn PM. Wilhelm Conrad Roentgen (1845-1923), the discovery of X-rays and perinatal diagnosis. Archives of Disease in Childhood. Fetal and Neonatal Edition. 2001;**84**(2): 138-139

[3] Eldek S. X-Ray Diffraction Lecture (Published in Education, Technology and Medicine). 2013. Available on https://www.slideshare.net/ouniyeh/x-ray-diff-lecture-3

[4] Oxtoby DW, Nachtrieb NH. Principles of Modern Chemistry. 3rd ed. Orlando: Saunders College Publishing; 1996. pp. 483-494

[5] Johanssen AE, Campbell JL, Malmqvist KG. Particle-induced X-ray emission spectrometry (PIXE). In: Chemical Analysis—A Series of Monographs on Analytical Chemistry and its Application. New York: John Wiley and Sons, Inc.; 1995. pp. 1-20

[6] Ebbing DD. General Chemistry. Massachusetts: Houghton Mifflin Company; 1984. p. 147

[7] Brouwer PN. Theory of XRF—Getting Acquainted with the Principles. 3rd ed. The Netherlands, 25: PANAlytical BV; 2010. p. 10

[8] PANalytical, The Analytical X-ray COMPANY, Brochure on X-ray Fluorescence and X-ray Diffraction

[9] ThermoFisher Scientific. XRF Technology in the Lab. 2017. https://thermofisher.com/content/sfs/brochures/TS-XRFLab-ebook-21Nov 14.pdf, [Retrieved on October 05, 2017]

[10] Elemental Analysis, Inc. Proton Induced X-ray Emission. 2016. http://www.elementalanalysis.com/pixe.html [Retrieved on August 05, 2017]

[11] Centre for Energy Research and Development (CERD), Obafemi Awolowo University, Ile-Ife, Nigeria, Monograph on Ion Beam Analysis (IBA) Techniques

[12] National Electrostatics Corp. 2017. http://www.pelletron.com/wpcontent/uploads/2017/02/Alphatross-v1.pdf [Retrieved on October 15, 2017]

[13] Igwebike-Ossi CD. Effects of combustion temperature and time on the physical and chemical properties of rice husk ash and its application as extender in Paints [PhD thesis]. Department of Pure and Industrial Chemistry, Univerity of Nigeria, Nsukka; 2011

[14] Igwebike-Ossi CD. Elemental analysis of rice husk ash using proton-induced X-ray emission (PIXE) spectrometry. International Journal of Applied Chemistry. 2016;**12**(3):233-242

[15] Igwebike-Ossi CD. Elemental analysis of rice husk using proton-induced X-ray emission spectrometry. International Journal of Applied Chemistry. 2017;**13**(4):801-811

[16] Igwebike-Ossi CD. Potassium oxide analysis in rice husk ash using proton-induced X-ray emission (PIXE) spectrometric technique. International Journal of Applied Chemistry. 2016;**12**(3):281-291

[17] Malcolm PS. Polymer Chemistry—An Introduction. New York, USA: Oxford University Press, Inc.; 1990. p. 140

[18] Housecroft CE, Alan GS. Inorganic Chemistry. 3rd ed. Essex, England: Pearson Education Limited; 2008. p. 166

[19] Mahan BM, Myers RJ. University Chemistry. 4th ed. California: The Benjamin Cummins Publishing Company, Inc.; 1987. pp. 998-1002

[20] Khandpur RS. Handbook of Analytical Instruments. 2nd ed. New Delhi, India: Tata McGraw Hill Education Private Ltd.; 2006. pp. 344-351

Chapter 11

Damage Detection and Critical Failure Prevention of Composites

Mark Bowkett and Kary Thanapalan

Additional information is available at the end of the chapter

http://dx.doi.org/10.5772/intechopen.71245

Abstract

In this chapter, critical failure prevention mechanism for composite material systems is investigated. This chapter introduces both non-destructive failure detection methods and live structural tests and its applications. The investigation begins by presenting a brief review and analysis of current non-destructive failure detection methods. The work proceeds to investigate novel live structural tests, tomography and applications of the proposed techniques.

Keywords: critical analysis, composites, smart materials, printed circuit board (PCB), damage detection

1. Introduction

Failure detection methods of carbon composite material systems are currently the subject of much research effort in the composite material community at large; see, for example [1–4], using a variety of failure detection methods and control algorithms. For enhanced reliability, early failure detection methods with critical failure prevention are preferable. Therefore, early failure detection techniques have become increasingly popular in the composite material systems community. Live failure detection techniques in composite material systems offer a structured approach to resolve failure-related issues giving essential early indication and warning. In this chapter, details of live failure detection techniques will be discussed in addition a brief review and analysis of current non-destructive failure detection methods of composite materials.

It is necessary to review current failure detection methods and determine if any of these methods have the potential to be used on low cost consumer products with scope for high end specialised equipment. This will determine if there is a potential technology gap that if filled

INTECH
open science | open minds

© 2017 The Author(s). Licensee InTech. This chapter is distributed under the terms of the Creative Commons Attribution License (http://creativecommons.org/licenses/by/3.0), which permits unrestricted use, distribution, and reproduction in any medium, provided the original work is properly cited. (cc) BY

will have a significant advantage to the consumer and specialist that can determine the state of health of equipment commonly in use. For this to be successful, the solution must be cost effective, robust, require low damage inspection knowledge and skills and be readily available.

There are a number of common causes for damage to occur and it can be certain that once there is a damage this will perpetuate further. The damage of a composite and its components can be attributed to different stages in their life: during manufacture, construction and the in-service life of the composite. A matrix crack typically occurs where there has been a high stress concentration or can be associated with thermal shrinkage during manufacture, especially with the more brittle high-temperature adhesives. Debonding occurs when an adhesive stops adhering to an adherend or substrate material. Debonding occurs if the physical, chemical or mechanical forces that hold the bond together are broken. Delamination is a failure in a laminate, often a composite, which leads to separation of the layers of reinforcement or plies. Delamination failure can be of several types, such as fracture within the adhesive or resin, fracture within the reinforcement or debonding of the resin from the reinforcement [3].

A void or blister is a pore that remains unoccupied in a composite material. A void is typically the result of an imperfection from the processing of the material and is generally deemed undesirable. Because a void is non-uniform in a composite material, it can affect the mechanical properties and lifespan [5]. Blisters are generated in the outermost layers. Porosity can be caused by volatile entrapment during the curing of the resin.

Wrinkles are common when adding new layers; it is significant to eliminate them as they can weaken the composite [6]. The inclusion of foreign bodies in the composites can include backing film, grease, dirt, hair and finger prints, which can lead to areas rich or deprived of resin [7]. To avoid the occurrence of a catastrophic failure due to manufacturing defects, impacts or fatigue damage, critical structural components are regularly inspected using various non-destructive testing methods.

The visual inspection is the most basic type of non-destructive testing method for composites [8]. Another quick and easy method for detecting exposed carbon fibres is to run a dry cloth over the surface as the fibres become easily snagged on damaged parts of the structure due to exposed fibres, this is immediately apparent to the inspector [9].

To increase the chances of visual inspection, a force may be applied to the structure acting as a manual flex test, which would further open up any cracks making it more likely to be seen. However, if proper care is not taken with this approach, it is possible that the carbon fibre can be excessively flexed beyond its damaged capabilities and incurs additional damages. The tap test is another simple test that can be performed as part of routine maintenance [10]. However, this technique is highly operator dependent as it requires a 'feel' for how it is meant to sound [11].

Radiography type testing uses X-ray and gamma rays for detecting internal imperfections or defects [12]. Ultrasonic inspection works by sending a high frequency sound wave into the structure and then measuring the reflected sound wave. The amount of energy transmitted or received and the time the energy received is analysed to determine the presence of flaws [13]. Dye penetrant method [4] can be used to detect the materials surface defects, but as it can only reveal surface defects and does not give details of depth of defects, it can also be difficult to test coarse surfaces. Pulse thermography [14] is an advanced non-destructive testing method, in this method thermal imaging cameras are used to detect material failures [15].

Acoustic method [12] is a structural health condition monitoring method which can be used for continuous monitoring of in-service structural components and help increase confidence regarding the remaining in-service lifetime if a fatigue limit cannot be defined easily. This method refers to the generation of transient elastic waves that are created by sudden redistribution of stress in a material. When the material is subjected to a change in pressure, load or temperature localised sources trigger the release of energy. The energy released is in the form of stress waves which propagate through the material and to the surface. It is possible to observe such stresses with suitable sensors mounted on the material. In composite materials, it is feasible to monitor for matrix cracks, fibre breaks and debonding.

Eddy current testing uses a circular current to detect the presence of cracks, surface breakings and variations in the composition of materials as well as identifying the material itself. It is an electromagnet testing which is one of the oldest testing methods [16]. However, its limitations are that only electrically conductive materials can be inspected, the surface must be accessible to the probe, an excellent level of inspector training and experience is required, rough finishes can interfere with the test, depth of penetration is limited and it is not suited towards large area testing.

A review of reported non-destructive testing methods for failure detection and prevention shows that many approaches require the composite structure either be taken to a test house or that relatively complex and large equipment be taken to the structure site [9]. In each case the equipment is large, requires a high level of competence and is typically expensive. Furthermore, the range of defects is wide and so requires advanced techniques to detect their presence, which leads to the development of live failure techniques in composite materials.

2. Design innovation

It is understood that there is a requirement in a relatively unexplored area that can be broadly classified as live failure techniques in composite materials. Continuing with carbon fibre as the material of interest, it is necessary to investigate various methods of resolving issues currently unsolved. Design innovation is intended to provide a structured approach to resolve such issues with clear and guided paths for which the theoretical solutions can be documented, analysed, assessed, researched and progressed through feasibility studies and ultimately development and prototyping of the product. Such an approach gives alternative direction and multiple concepts should a particular theoretical solution have a shortcoming or worst case fail to deliver on its targets. With several theoretical concept solutions documented and assessed for practical feasibility, we can continue to progress into a practical environment and begin the early stages of product development.

It is prudent in such an application as critical failure of composites to have a reference point in which to determine the successfulness of a proposed live failure system. For this reason, the first step before commencing works on a practical solution is to develop and document working procedures for the creation of test subjects or specimens, in this case small samples (batches) of carbon fibre specimens. Progressing on with this methodology, similar procedures were again developed and documented for stress testing carbon sample specimens. This gives a solid foundation in the form of controlled test specimens along with quantised data in which implementations of the theoretical concept solutions can be applied and importantly evaluated against and thus measured for success.

2.1. Control structure

The control specimens comprise of a small strip of 430 g 2/2 twill vari preg carbon fibre with the dimension 30 × 300 mm, these are four layer plies of identical ply orientation. The samples are cured in a preheated environmental chamber set at 100°C for 90 min which conforms to the guidelines on the carbon fibre vari preg data sheet supplied by the manufacturer. The control structure is set at these parameters as it allows for relatively fast builds due to the low ply count and is of suitable size for structural tests, the low dimensions also gives minimal manual labour when applying the proposed theoretical concept solutions for the failure detection methods already set out in the design innovation phase. It is essential that such rapid builds are possible due to the necessity to test destruction for each sample specimen in order to observe its behaviour when no foreign bodies are included. This is an essential requirement as additions that will be added later such as sensors or probes, for example, must not compromise the structural integrity of the composite which could actually lead to a lower strength composite material. To reiterate, it is considered desirable that any proposed early detection method do not impose a penalty in terms of the composites structural strength as it would be prior to the application of the detection system.

A small control sample batch of 10 units (specimens) is produced and logged, the higher the quantity of the control sample batch will increase the comparison accuracy of subsequent batches that are fitted with the failure detection system. The exact number of batch quantities lies with the project size and desired test plans. In this case, it was deemed more appropriate to run smaller samples during the initial concept phase of the project where the failure detection method was likely to be frequently altered. When the failure detection system is proved suitable for its intended operation at a more mature time during its development, then the batch size will be increased as only refinements will be necessary at this stage. This scaled approach allows for rapid prototyping that leads onto a slower polished version as the direction becomes more apparent from the research activities.

Each individual control sample specimen was analysed and the data logged. Flexural tests on each sample were conducted using the universal testing machine (see **Figure 1**). In the image, it can be seen that we have two points supporting the control specimen of carbon fibre and epoxy resin composite, the third point is applied from above and applies the deformation on the sample. The test is automatically monitored by a personal computer that records the deformation distance in millimetres and the force in Newton, from this a graph can be plotted.

Individual data analysis logs details such as the sample specimen weight. The mean values, variance and standard deviation are calculated with the accompanying formula. This gives

Figure 1. Flexural test equipment: universal testing machine.

a quick to view reference point that can be easily remembered and referenced against for future builds that utilise a damage detection technique and can give indication to how much the damage detection system has altered the properties of the native carbon fibre composite.

Figure 2 shows key characteristics of the control sample specimens tested with the flexural test setup on the universal testing machine. Collective data results of the three point flexural test are shown in **Figure 3**.

Referring back to **Figure 2**, the graph shows the data recorded for specimen 8 of the control batch with regions of interest highlighted. It can be seen that the sample initially operates in its linear and elastic range as expected. The first potential sign of damage is at a deformation of approximately 16 mm and this can be seen as 'elastic but no longer linear' on the graph. This first sign of damage and could be heard as a low volume audible crack from the carbon fibre. Under visual observations absolutely no damage was observed. Increasing deformation to 19 mm equating to 215 N, a second crack is heard with increased volume and a noticeable movement in the carbon fibre was observed as the curvature of the sample marginally straightened from the test point. At this point, visual observations failed to recognise actual surface damage, it is assumed, however, that removing the specimen from the test machine at this point would reveal a permanent deformation in the specimen that was originally entirely flat. From this deformation point onwards appreciable crackling could be heard as the matrix and reinforcement broke down. It was only at the fracture point that it was apparent that the composite had failed but was still bound by the fibres in such a way that it was still an intact single piece of carbon composite. Ideally, a failure detection technique would be able to monitor damage when the material has gone beyond its linear and elastic range as this would give the maximum possible time to the user that the composite was approaching failure. Due the sensitivity required to detect this minuscule disturbance (which was not present across all control sample specimens tested), it is considered more appropriate to detect the region of yield strength where stronger indications of damage occur. It is assumed that if an alert can

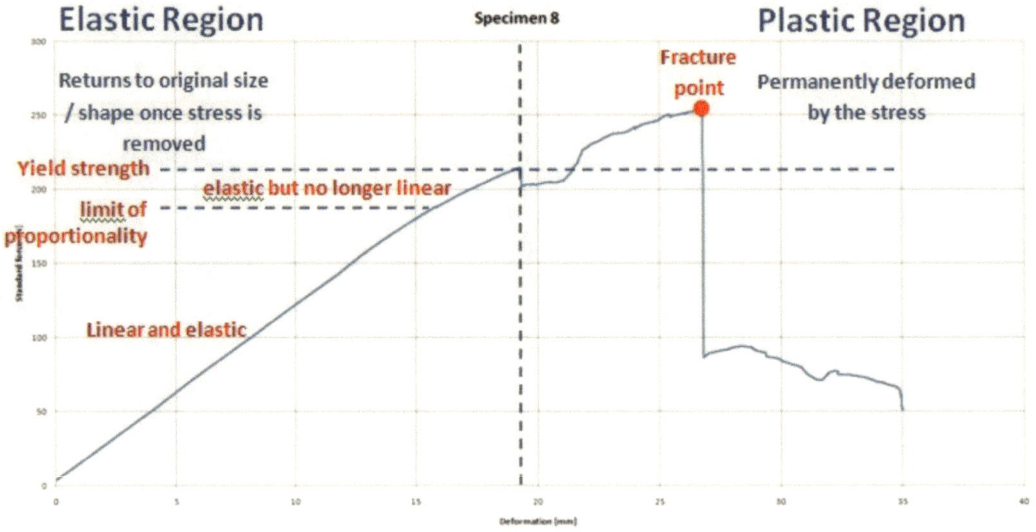

Figure 2. Control sample specimen.

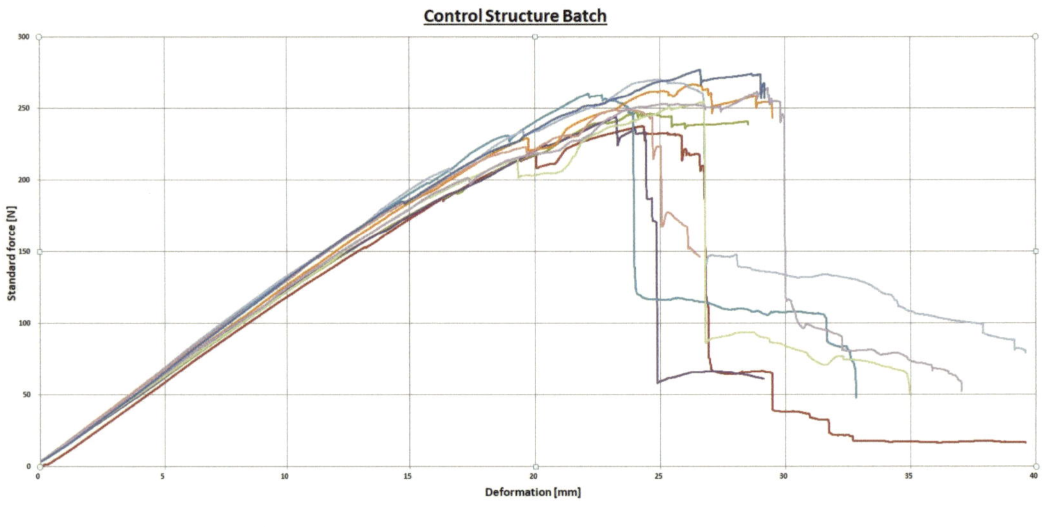

Figure 3. Collective data of three point flexural test.

be issued to the operator of the structure (e.g. bicycle) at this point that enough prior warning has been given to be able to come to a safe stop and have the structure assessed by more in depth damage detection equipment such as X-ray.

These issues occur in many different applications. In this section two examples are discussed, firstly a bicycle application and then a quadcopter application.

Graphs are recorded and analysed on an individual basis but can be seen collectively (see **Figure 3**) for the three point flexural test: The Y-axis represents the force in Newton and the

X-axis is the deformation in millimetres. Observations show that each sample follows a similar trend but can vary at its key parameters such as fracture point, etc.

The mean weight of each sample is 26.37 g and this is a good indicator to the amount of material used in the sample (carbon fibre and epoxy resin), it can be helpful to reference this as to gauge consistency of the manual work during the construction of the sample specimens. Other parameters and their associated equations can be seen below:

- Weight (variance):

$$\sigma^2 = \frac{\sum x^2 - \frac{(\sum x)^2}{N}}{N} = 0.15\,g \quad (1)$$

Control specimen weight variance of 0.15 g from a mean value of 26.37 g.

- Peak force (variance):

$$\sigma^2 = \frac{\sum x^2 - \frac{(\sum x)^2}{N}}{N} = 133.8\text{ N} \quad (2)$$

Control specimen peak force variance of 133.8 N from a mean value of 255.9 N.

- Yield strength (variance):

$$\sigma^2 = \frac{\sum x^2 - \frac{(\sum x)^2}{N}}{N} = 40.41\text{ GPa} \quad (3)$$

Control specimen yield strength variance of 40.41 GPa from a mean value of 192.3 GPa.

Although bicycle applications are relatively safe, as they categorised in the land section of applications, many serious accidents do occur from continued use as the rider is unaware that the structure has been pushed past its performance envelope. Life-threatening events are more likely to occur in aviation structures such as planes if this were to go undetected. Although arguably impossible to come to a gradual stop in such a situation, if the pilot were alerted to such detection it would be possible to 'limp home', where by the aircraft would be restricted to low G movements such as turns or deceleration. Unexpectedly similar risks can be expected in unmanned aircraft or the ever increasingly popular quadcopters or drones. Although no immediate threat of life is assumed due to the lack of an onboard pilot, drones are increasingly flown in areas of large crowds due to their ability to carry high end photography equipment. It is no longer uncommon for higher end drones to approach 10 kg in weight and exceed this, due to large professional cameras for photography and film industry. It is therefore appreciated that the risk of life would be to the crowds immediately below should damage be undetected to one of the motor arms resulting in a complete lack of vertical thrust.

2.2. Mesh structure

To satisfy the requirements of live failure detection systems at its most basic level was to incorporate what was deemed as the simplest concept solution being the 'mesh structure concept'. For this two options are available, in the first instance a simple conductive mesh with insulating material is embedded within the carbon plies (**Figure 5**, left image), this thin diameter mesh was constructed of low gauge enamelled copper wire with a diameter of 0.22 mm and applied to the inner plies of the carbon fibre stack before curing. The mesh wires are allowed to extrude from the carbon fibre as flying leads from which simple test equipment can be attached such as a multimeter. Currently, the mesh is created from a single piece of wire which gives two open-ended flying leads, this offers the most simple and rapid technique to embed the mesh for research purposes. The mesh, however, is not limited to this single wire as it is possible to use multiple wires with the advantage of a means of simple damage location, however, this introduces greater complexity and additional electronic hardware to monitor the system, it is still an uncomplicated method. Arguably the requirement for multiple wire systems are not essential for simple carbon fibre constructs nevertheless it does give a factor of flexibility should the design constraints demand more precise failure detection which requires location data.

Details of the different applications in simple and complex structures will be explained later. However, in order to explain the mesh structure in detail, in this section, the quadcopter application within the aeronautical sector (see **Table 1**) is considered.

The mesh structure can be more easily understood with reference to the quadcopter CAD diagram (see, **Figure 4**, right image). In this case, the quadcopter frame is constructed of glass reinforced epoxy laminate (FR4) more commonly used in printed circuit board (PCB) manufacturing. It is a composite material comprised of a flame resistant [17, 18] woven fibreglass cloth and an epoxy resin binder. It can be seen that the front half of the quadcopter frame (upper area) has no failure detection system incorporated, whereas the rear half (lower area) has the basic level of failure detection integrated onto the FR4 board (see **Figure 4**). This simply includes a single track of copper at 1 oz. which equates to an approximate thickness of 0.089 mm. At the rear (centre lower area) of the frame are two pads in which suitable electronics can be connected in order to monitor that the wire mesh has not gone open circuit as a result of physical damage such as a fracture, for example. This monitoring signal can be fed to the flight controller and transmitted to the user via flight telemetry data. It is almost effortless to separate the wiring for each arm if damage location is to be realised, giving an adequate enhancement if the user requires data as to know which quadcopter arm has sustained physical damaged. The diagram shows the failure detection method as a red line (copper PCB trace), this has been applied to the upper layer of the PCB to allow for a visual demonstration

Catagory	Example 1	Example 2	Example 3	Example 4
Aeronautical	Aircraft	Aerospace	Unmanned ariel vehicles	Drone taxi
Land	Formula 1	Cars	Bikes	Composite wheels
Nautical	Submersibles	Hulls	Yacht masts	Booms
Leisure	Bicycles	Golf clubs	Windsurf masts	Rackets
Business	Security	Dispatch	Wind turbine blades	Architecture
Military	Combat vehicles	Armoured vehicles	Essential electronics	Fighter jets

Table 1. Application list examples.

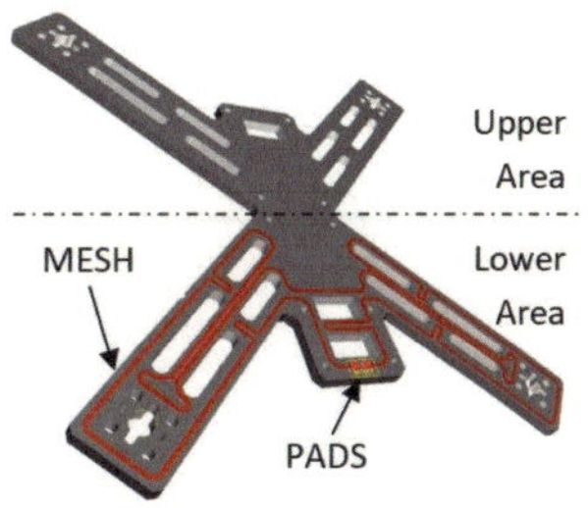

Figure 4. Traditional drone frame (left), enhanced NDT CAD frame (right).

of the system, however, it is possible to incorporate this to the inner layers or bottom of the board as desired by the designer. It is easy to realise the simplicity and the benefits of this approach especially when compared to the available quadcopter frames currently on the market as shown in **Figure 4** (left image). Here the veined arms reduce weight but shows obvious risks of catastrophic failure should a single element be damaged and go undetected.

PCBs are readily available in various thicknesses, multiple materials and layer makeup's offering an applicable solution for a variety of applications. Increased thickness of the FR4 board improves rigidity whilst a lower thickness improves flexibility allowing for lower FR4 thicknesses of 0.4 mm to be curved around existing structures such as carbon fibre. It should be noted that appropriate adhesion be applied spanning the entire FR4 board as poor contact can allow fractures in the hosts structural material not to propagate to the FR4 failure detection board. Additional precautions should be noted as the addition of two different composites simply stuck together brings potential problems due to differing mechanical properties inherent with the constituent composites. As an example, the Young's modulus of standard carbon fibre is 70 GPa where FR4 is 24 GPa, similarly thermal expansion coefficient variance would be of concern at temperature ranges if the individual composites were not suitably decoupled. It may be considered suitable in certain situations and this is left to the engineer to utilise appropriate combinations of composites for the environment and that of the host structure.

In the second instance, the wire mesh can be added as an aftermarket product to existing carbon fibre structures or even non-conductive structures such as fibre glass. This would typically be applied as a single unit, fixing a mesh as a single wire to structures can be labour intensive and cumbersome. It is, therefore, more appropriate to have the mesh incorporated on an adhesive sticker and applied by normal manufacturing routes.

The benefit of the mesh structure is that the detection electronic hardware is extremely simple, requires very low real estate of the host structure and its operational power consumption is almost negligible, lending itself perfectly to long-lasting portable applications. Further, more such a system can be powered by energy harvesting methods such as vibrations, solar, wind and the like, this will obviously incur additional constraints in terms of size and cost of the overall damage detection product. The detection principle is a simple case of measuring current flow through the conductive copper mesh, when damage occurs as a result of a crack or over flex in the structure then the conductive wire is severed ceasing current flow, allowing the user to be alerted to the fault. Such a simple solution has its draw backs and this is the location accuracy to the damaged area. In preliminary lab tests on wire mesh test specimens only ~50% of flexural test fractures were detected before a catastrophic failure event. Analysis shows that the reason for this was down to one of two reasons; either the mesh wire was not present in the fracture line or that the fracture width was not great enough to be detected. The image (see **Figure 5**) is taken at ×200 magnification of a mesh structure specimen:

Observations show that the 0.22 mm enamelled copper wire has stretched with the fracture during flexural testing, ideally, this would have sheared and broke at the same rate as the carbon fibre. To improve the system, it is suggested that the detection material to have a similar Young's modulus to that of the material under test and that a suitable pitch be used for the mesh to be fitting to that of the application. However, this method has proved an extremely low cost and portable method for additional safety where there was none. The application has been used in low cost multirotor (quadcopter) frames in particular the motor arms where damage could be incurred from in-flight collisions such as trees, buildings and the like. This gives an entry level of security against further damage should such a collision occur and the operator continue to fly the multirotor, as without being immediately able to inspect for damage such damage is unknown.

2.3. Wafer structure

The sheet structure method utilises an insulating material as alternating ply layers with the carbon fibre plies to make the stack. In this concept specimen absorbent glass mat (AGM) is implemented which originates in lead acid batteries [19]. The AGM is reduced in thickness to a depth of approximately 1 mm and is sandwiched between the pre preg carbon fibre sheets. For example, the ply make up would consist of a first ply being carbon fibre, the second ply AGM, the third ply carbon fibre, and so forth until the desired layer make up is achieved. The theoretical objective behind this makeup is to monitor electrical conductivity between the now mostly electrically isolated plies of alternating carbon fibre, such as layer one and three, to continue the example mentioned above. When the alternating ply stack of carbon fibre and fibre glass is damaged the strands break and combine into a fibrous blend

Figure 5. Tapering of conductive mesh (up to ×200 magnification).

causing measureable electrical conductivity between the pre damaged isolated plies, therefore can be measured with simple portable electronic equipment. In practice the layers are not 100% isolated and the AGM caused the structure to fail more rapidly during flexural testing. It is predicted that longer glass fibres be used as the insulating material to provide increased structural integrity, or an alternative insulating material be used such as Kevlar, consideration should be given to Kevlar as it is absorbent to water. Caution should be taken as to the expansion properties of each constituent material and suitable proof of concept works is required to ensure the hybrid composite performs as desired. It is fair to assume that this approach would be a trade off or compromise to a pure carbon fibre composite. However at this time full laboratory testing has not been completed and the full benefits or disbenefits of such an approach is unproven, it is possible that other mixtures of plies can offer significant advantages over that of carbon fibre alone for certain applications such as fire retardant improvements.

3. Experimental study

Current research takes onboard a known technique that has been practiced for many years, possibly the earliest is the X-ray computed tomography, well known as the CT or CAT scan [20]. Similarly, a geophysical monitoring method known as electrical resistivity imaging or ERI typically uses four equidistant electrodes such as metal probes that are staked into the ground. Two of the four probes are used to induce a low frequency or direct current into the earth and the remaining two probes are voltage or potential probes which measure the resultant voltage potential. The voltage is simply converted into a resistance value that represents an average resistance between the probes, this method is often found in archaeology [21].

Another approach that is widely researched and utilised in the medical field is known as electrical impedance tomography (EIT) [22–23], surface electrodes usually physically organised in a ring layout are in direct contact with the skin, generally with electrocardiogram (ECG) electrodes and gel to improve conductivity. This is then positioned, for example, around the chest or leg and a small current is passed through the ECG probes whilst the passive ECG probes measure resistance and reactance, similar to other tomography techniques. The resultant post-processed image represents the phase shift in the measured signal depending on the technique used. The probe count is typically higher in comparison to other fields such as geological tomography to provide an increase in resolution of the image, 16 electrodes and higher is not uncommon. When a measurement has been taken electrodes are rotated until the full electrode array has been sampled, the resulting data is processed and the cycle begins again.

There are many types of electrical tomography but all are based on the same principle. The intention of which is to construct an image from a physical object of its observed measurements. As such this is an inverse problem and is one of the most important and well-studied mathematics problems in science and mathematics today [24]. The direct problem is relatively undemanding, the inverse one is particularly difficult.

Any solution to an inverse problem requires an element of estimation to fill in data that is not available and as such any solution cannot be void of resolution error or may be low of resolution. However, such techniques do provide suitable representations of the object in question and is a technological step forward.

With reference to **Figure 6**, the left most image labelled 'carbon fibre' shows the physical test setup of the embedded probes (electrodes) and the highlighted damaged area shown within red circle, where as the right image 'EIT image' shows a typical result from post-processing. It may be difficult to see damage to the physical carbon fibre structure visually due to the surface ply remaining intact; the EIT technique can also provide useful insight to the structural health of the composite.

Although the technique provides a solution for carbon fibre, various problems need to be addressed in order to deliver a suitable end product for mobile live applications. Consequently, fast processing of data is required with little computational overhead whilst maintaining suitable resolution for adequate damage detection. In addition omissions need to be made for any flex in the structure as this varies the structures electrical resistivity and can be mistakenly characterised as damage if the implemented algorithms are not accurate. Further exacerbating the situation, environmental conditions play a large role, temperature changes are picked up by the same technique used to detect damage due to the temperature coefficient of the composite material. The resultant solution to the problem of creating a portable live damage detection system must address each mentioned issue successfully if such an inverse problem is to be of practical use.

However, techniques such as tomography is widely in use today within various scientific fields and do offer a solution to where there was none. The additional problems mentioned such as resistance changes due to flex and the environment such as rain and sunlight causing localised temperature changes can be addressed with hardware or software implementations. At its fundamental level these challenges ultimately reduce the sensitivity of the tomography technique in composites and its effectiveness of detecting structural health conditions. As such, future research is required and ultimately real world testing of hardware and software algorithms to determine if this method is adequate across all applications.

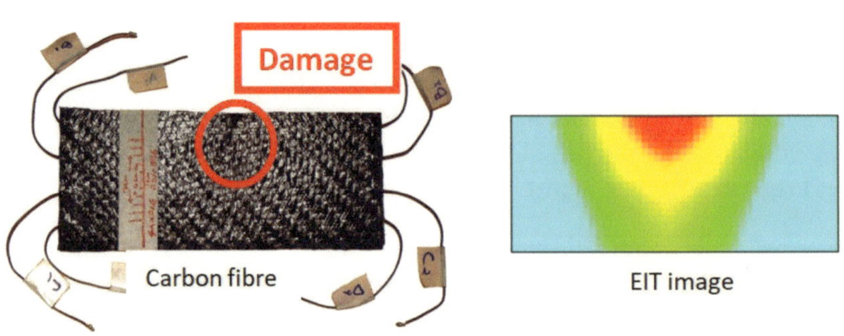

Figure 6. Test specimen & electrodes (left), expected EIT image (right).

4. Applications

There are many applications in which the suggested damage detection methods mentioned can be utilised in which no prior solution has been achieved other than offline methods, such offline methods require the structure in question be taken to a test laboratory for a skilled professional to access. Although the suggested methods are still progressing through the research phase and have yet to be formalised, such as production methods and materials for the mesh method, wafer method and the experimental method based on tomography technique, **Table 1** shows a list of potential applications for the concept techniques described.

The potential applications for online damage detection are arguably limitless and the table demonstrates the range of applications to illustrate this. The proposed concept methods include both forward and inverse problems of the mesh and tomography techniques, respectively. This allows a suitable system to be used depending on the requirements. Some of the less obvious applications mentioned that have not been covered already are discussed for clarification and completeness.

For the application of military damage detection, is it not uncommon for vehicles and associated parts to come under gun fire or explosive blasts, it is difficult if not impossible to assess the damage from within the relative safety of the vehicle. Although the vehicles are generally not constructed of carbon fibre but of metals or composites such as steel and ceramic it is possible to address the issue with the mesh technique. Thin sheets of mesh can be attached in suitable key locations behind the armour plating for detection of a breach in the vehicles structure. This would be of particular importance if essential drive train components or tactical equipment were present at this location. To enhance the detection further printed circuit boards can have an additional layer within the FR4 substrate or as an addition to the top or bottom layers that can act as the mesh damage detection, similar to the quadcopter CAD diagram (see **Figure 4**). This can give essential feedback to the vehicle users should an electronic system fail due to projectile damage from such gunfire or explosive devices.

5. Discussion and concluding remarks

In this chapter, a brief overview of carbon fibre has been given along with its common defects and the conventional non-destructive testing techniques. This basic ground work establishes the direction of current research and why the concept methods were chosen. This progresses to a simple overview of the design innovation approach and initial testing stages and comparison methods involving control specimens. A table of its intended applications are given to give an appreciation of the flexibility of the proposed end products that can be developed as a direct research outputs. To date there are no suitable live detection techniques available that can provide a portable solution for low cost consumer products or high end equipment in an easy to install manner.

The design innovation and research approach taken in this study has allowed for the potential development of flexible methodologies and products. The methodology as demonstrated in the mesh technique on the quadcopter may be taken immediately into existing designs such as PCBs for physical damage detection at a cost of adding an existing layer to the design, alternatively this can be routed into the existing layer should it be suitable. An advanced understanding of PCB design may be desirable as the effects of broken ground planes and radio frequency wave interference could result if the mesh technique is not correctly implemented. Any one skilled in the art can adopt such a methodology without adverse effects to the original design. This most basic level of damage detection is extremely low cost and simple to implement, the simplicity allows for a robust detection system that requires no processing power and is considered mechanical in its design. There are various ways in which this method can be practiced and the only limitation is that of the designer. Research is still continuing in this area to expand upon the range of applications, its successfulness in detecting damage and the operating margin at which the damage is detected.

Increasing in technological difficulty is the wafer technique which relies upon alternating conductive and insulating plies. This is another mechanical method and is a forward problem like the mesh concept, therefore offering a simplistic solution and inherent robustness within the design. Although the works are in its early stages it can be appreciated that only certain materials may be of interest for this to be successful. For example, it is necessary for a conductive ply material and non-conductive ply material to make up the stack, at minimal a three ply stack would be necessary. Unlike conventional composites where there is typically only one reinforcement material there would now be two but the matrix in current research motioned in the text has been kept at one and is a polymer resin. It is worth mentioning that this is not considered a hybrid material as these are composites commonly using one organic and one inorganic compound at the nanometre or molecular level. However, this does bring an alternative possibility and research route for hybrid material design specifically targeted for damage detection. The undersized research effort concentrated on this technique has been investigated on traditional carbon fibre and an experimental material of AGM. As previously mentioned in the chapter, the relatively short fibres of the AGM had such a negative effect on the strength of the structure that no further tests were carried out. Research efforts will continue in this area on traditional fibre glass sheets as used in composites to increase structural integrity. It can be appreciated that this will not be as strong if only carbon fibre was used as a single material but this would be a known sacrifice for the damage detection system should it prove adequate.

Finally, the experimental technique involves the tomography approach, although this technique is approached by many in the research community there is yet to be of any commercial products available that exploits its use in commercial composite structures. This approach is considered advanced due to the inverse problem and potentially difficult to implement in practical applications. In lab tests in a controlled environment, it is relatively simple to implement and offers suitable detection of damage, bench tests were conducted on physically small specimens with advanced equipment. The damage detection values are of such sensitive levels that it may be difficult to engineer the electronic circuitry required for portable applications requiring low power consumption with today's technology. Under the bench test conditions heat changes to the composite due to manual handling were detected as damage, this was also true of temperature changes from air conditioning units being switched on or off. The effectiveness of this method can never be absolute due to the inverse problem and this is exacerbated by the need to mask temperature drift and sudden changes of the composite. The actual value of this technique as such will not be realised until fully inspected.

Collectively a broad spectrum of novel techniques and methods have been conceptualised within the three categories mentioned in this chapter with a comprehensively range of technical difficulty. Continued research is required to prove the effectiveness in practical applications and can only be realised with future efforts. The ultimate goal of which is to develop a new range of products and methods that can give early warning and/or locations of physically damaged structures or components that comprise of systems that would cause further injury or critical failure if left undetected in use.

Author details

Mark Bowkett* and Kary Thanapalan

*Address all correspondence to: mark.bowkett@southwales.ac.uk

Faculty of Computing, Engineering and Science, University of South Wales, UK

References

[1] Cheng L, Tian GY. Comparison of non-destructive testing methods on detection of delaminations in composites. Journal of Sensors. 2012;**2012**:1-7. Article ID: 408437

[2] Luo Y, Wang Z, Wei G, Alsaadi FE, Hayat T. State estimation for a class of artificial neural networks with stochastically corrupted measurements under round-robin protocol. Neural Networks. 2016;**77**:70-79

[3] Unnporsson R, Jonsson MT, Runarson TP. NDT Methods for Evaluating Carbon Fiber Composites. Bristol: CompTest, Bristol University; 2004

[4] Bowkett M, Thanapalan K. Comparative analysis of failure detection methods of composites materials systems. Systems Science & Control Engineering: An Open Access Journal. 2017;**5**(1):168-177

[5] McEvoy MA, Correll N. Materials that couple sensing, actuation, computation, and communication. Science. 2015;**347**(6228):1261689

[6] Zhang Z, Hartwig G. Relation of damping and fatigue damage of unidirectional fiber composites. International Journal of Fatigue. 2002;**24**(7):713-718

[7] Adams RD, Cawley P. Review of defect types and non-destructive testing techniques for composites and bonded joints. NDT International. 1988;**21**(4):208-222

[8] Gholizadeh S. A review of non-destructive testing methods of composite materials. Procedia Structural Integrity. 2016;**1**:50-57

[9] Bowkett M, Thanapalan K, Williams J. Review and analysis of failure detection methods of composites materials systems. Proceedings of the 22nd International Conference on Automation & Computing, Colchester, UK, IEEE; 2016. pp. 138-143

[10] Smith RA. Composite defects and their detection. Materials Science and Engineering. 2009;**5**(3):103-143

[11] Greene E. Marine composites non-destructive evaluation. Ship Structure. 2014;**1**:416-427

[12] Ning W. Structural Health Condition Monitoring of Carbon-Fiber Based Composite Materials Using Acoustic Emission Techniques. MSc thesis. UK: College of Engineering and Physical Sciences, University of Birmingham; January 2015

[13] Li Z, Haigh AD, Soutis C, Gibson AAP, Sloan R, Karimain N. Damage evaluation of carbon-fiber reinforced polymer composites using electromagnetic coupled spiral inductors. Advanced Composites Letters. 2015;**24**(3):44-47

[14] He Y, Tian G, Pan M, Chen D. Impact evaluation in carbon fiber reinforced plastic (CFRP) laminates using eddy current pulsed thermography. Composite Structures. 2014;**109**:1-7

[15] Sharath D, Menaka M, Venkatraman B. Defect characterization using pulsed thermography. Journal of Nondestructive Evaluation. 2013;**32**(2):132-141

[16] De Goeje MP, Wapenaar KED. Non-destructive inspection of carbon fiber-reinforced plastics using eddy current methods. Composites. 1992;**23**(3):147-157

[17] Sarvar F, Poole NJ, Witting PA. PCB glass-fiber laminates: Thermal conductivity measurements and their effect on simulation. Journal of Electronic Materials. 1990; **19**(12):1345-1350

[18] Azar K, Graebner JE. Experimental determination of thermal conductivity of printed wiring boards. Proceedings of the Twelfth IEEE SEMI-THERM Symposium: 169-182, 1996

[19] Mariani A, Thanapalan K, Stevenson P, Williams J. An advanced prediction mechanism to analyse pore geometry shapes and identification of blocking effect in VRLA battery system. International Journal of Automation and Computing. 2017;**14**(1):21-32

[20] Du Plessis A, Le Roux SG, Guelpa A. Comparison of medical and industrial X-ray computed tomography for non-destructive testing. Case Studies in Nondestructive Testing and Evaluation. 2016;**6**:17-25

[21] Fry RJ. Time-Lapse Geophysical Investigations over Known Archaeological Features using Electrical Resistivity Imaging and Earth Resistance, PhD thesis. UK: School of Life Sciences, University of Bradford; 2014

[22] Brown BH. Electrical impedance tomography (EIT)—A review. Journal of Medical Engineering & Technology. 2003;**27**(3):97-108

[23] Bodenstein M, David M, Markstaller K. Principles of electrical impedance tomography and its clinical application. Critical Care Medicine. 2009;**37**(2):713-724

[24] Aster R, Borchers B, Thurber C. Parameter Estimation and Inverse Problems. 2nd ed. Cambridge, Massachusetts: Elsevier; 2012

Chapter 12

Fracture Variation of Welded Joints at Various Temperatures in Liquid-Phase-Pulse-Impact Diffusion Welding of Particle Reinforcement Aluminum Matrix Composites

Kelvii Wei Guo

Additional information is available at the end of the chapter

http://dx.doi.org/10.5772/intechopen.71249

Abstract

The fracture variation of liquid-phase-pulse-impact diffusion welding (LPPIDW) welded joints of aluminum matrix composites (ACMs: SiC_p/A356, SiC_p/6061Al, and Al_2O_{3p}/6061Al) was investigated. Results show that under the effect of pulse-impact (i) initial pernicious contact state of reinforcement particles changes from reinforcement (SiC, Al_2O_3)/reinforcement (SiC, Al_2O_3) to reinforcement (SiC, Al_2O_3)/matrix/reinforcement (SiC, Al_2O_3) and (ii) the fracture of welded joints with optimal processing parameters is the dimple fracture. Meanwhile, scanning electron microscope (SEM) of the fracture surface shows some reinforcement particles (SiC, Al_2O_3) in the dimples. Moreover, the slight reaction occurs at the interfaces of SiCp/6061Al, which is propitious to improve the property of welded joints because of the release of internal stress caused by the hetero-matches between the reinforcements and matrix. Consequently, aluminum matrix composites (SiC_p/A356, SiC_p/6061Al, and Al_2O_{3p}/6061Al) were welded successfully.

Keywords: aluminum matrix composite, fracture, fractography, particle reinforcement, pulse-impact, diffusion welding

1. Introduction

Aluminum matrix composites (AMCs) have a wide application in the fields of aerospace, automobile, structural components, heat-resistant-wearable parts in engines, and so on due to their high specific strength, rigidity, wear resistance, corrosion, and good dimensional stability [1–5]. The reinforcements in AMCs may be either in the form of particulates or as short fibers, whiskers, etc. [5, 6]. These discontinuous natures create several problems to their joining techniques for acquiring their high strength and good quality welded joints.

INTECH
open science | open minds

© 2017 The Author(s). Licensee InTech. This chapter is distributed under the terms of the Creative Commons Attribution License (http://creativecommons.org/licenses/by/3.0), which permits unrestricted use, distribution, and reproduction in any medium, provided the original work is properly cited. (cc) BY

The high specific strength, good wearability, and corrosion resistance of aluminum matrix composites (AMCs) attract substantial industrial applications. Typically, AMCs are currently used widely in automobile and aerospace industries, structural components, heat-resistant-wearable parts in engines, etc. [7–10]. The particles of reinforcement elements in AMCs may be either in the form of particulates or as short fibers, whiskers, and so forth [10, 11]. These discontinuous natures create several problems to their joining techniques for acquiring their high strength and good quality welded joints. Typical quality problems of those welding techniques currently available for joining AMCs [12–20] are as elaborated below.

(1) The distribution of particulate reinforcements in the weld.

As properties of welded joints are usually influenced directly by the distribution of particulate reinforcements in the weld, their uniform distribution in the weld is likely to give tensile strength higher than 70–80% of the parent AMCs. Conglomeration distribution or the absence (viz., no reinforcement zone) of the particulate reinforcements in the weld generally degrades markedly the joint properties and subsequently resulted in the failure of welding.

(2) The interface between the particulate reinforcements and aluminum matrix.

High welding temperature in the fusion welding methods (typically TIG, laser welding, electron beam, etc.) is likely to yield pernicious Al_4C_3 phase in the interface. Long welding time (e.g., several days in certain occasions) in the solid-state welding methods (such as diffusion welding) normally leads to (i) low efficiency and (ii) formation of harmful and brittle intermetallic compounds in the interface.

To alleviate these problems incurred by the available welding processes for welding AMCs, a liquid-phase-pulse-impact diffusion welding (LPPIDW) technique has been developed [21–23]. This work aims at providing some specific studies that influence the pulse-impact on the fractography variation of welded joints. Analysis by means of scanning electron microscope (SEM), transmission electron microscope (TEM), and X-ray diffraction (XRD) allows the micro-viewpoint of the effect of pulse-impact on LPPIDW to be explored in more detail. Also, temperature distribution in heated sample also calculates.

2. Experimental material and procedure

2.1. Material

Stir-cast SiC_p/A356, P/M SiC_p/6061Al, and Al_2O_{3p}/6061Al aluminum matrix composite, reinforced with 20%, 15% volume fraction SiC, Al_2O_3 particulate of 12 μm, 5 μm mean size, were illustrated in **Figure 1**.

2.2. Experimental procedure

The quench-hardened layer and oxides induced by wire-cut process, on the surfaces of aluminum matrix composite specimens, were removed by polishing with 400 # grinding paper carefully. The polished specimens were then properly cleaned by acetone and pure ethyl alcohol so as to remove any contaminants off its surfaces. A DSI Gleeble®-1500D

http://dx.doi.org/10.5772/intechopen.71249

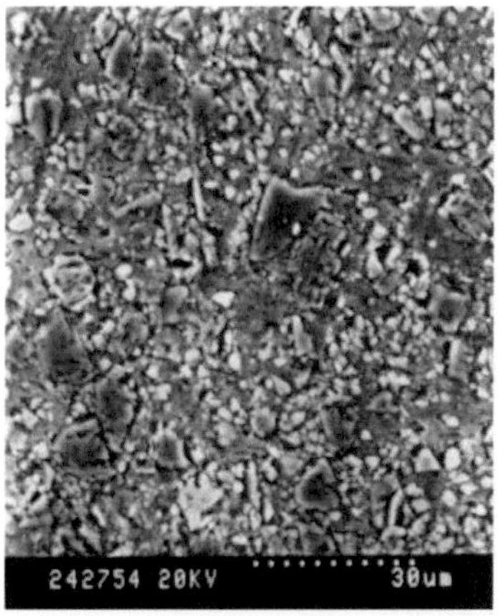

$SiC_p/A356$

SiC_p/6061Al

Al_2O_{3p}/6061Al

Figure 1. Microstructure of aluminum matrix composites.

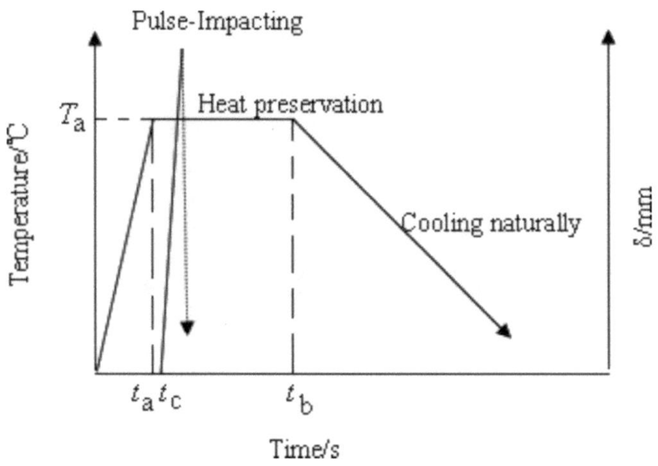

Figure 2. Schematic diagram of liquid-phase-pulse-impact diffusion welding.

thermal/mechanical simulator with a 4×10^{-1} Pa vacuum chamber was subsequently used to perform the welding.

The microstructures and the interface between the reinforcement particle and the matrix of the welded joints were analyzed by SEM, TEM, and XRD.

2.3. Operation of LPPIDW

Figure 2 shows a typical temperature and welding time cycle of a LPPIDW. It basically involves with (i) an initially rapid increase of weld specimens, within a time of t_a, to an optimal temperature T_a at which heat was preserved constantly at T_a for a period of (t_b-t_a); (ii) at time t_c, a quick application of pulse-impact to compress the welding specimens to accomplish an anticipated deformation δ within 10^{-4}–10^{-2} s, whilst the heat preservation was still maintained at the operational temperature T_a; and (iii) a period of natural cooling to room temperature after time t_b.

3. Results and discussion

3.1. Temperature distribution calculation in heated sample

The schematic diagram of temperature distribution in heated sample is shown in **Figure 3**, where L is the length of sample between two holders.

T_0 is the initial temperature. When the sample is heated, the heat transfer at the distance of x in Δt is

$$Q = -\alpha A \frac{\partial T}{\partial x} \Delta t \quad (1)$$

where α is the coefficient of conduction and A is the cross-sectional area.

http://dx.doi.org/10.5772/intechopen.71249

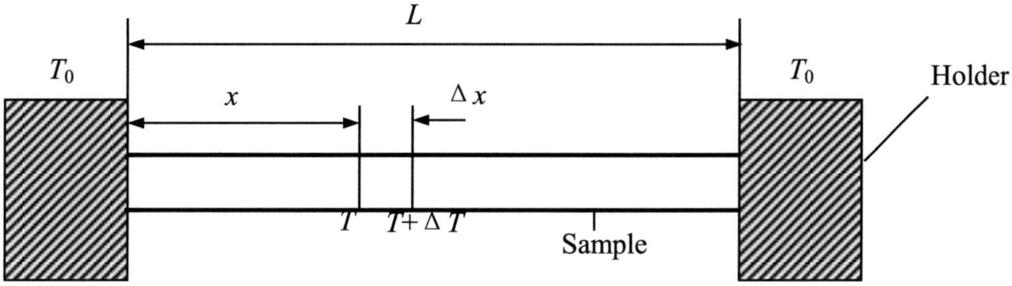

Figure 3. Schematic diagram of temperature distribution in heated sample.

At the same time, the heat transfer at the distance of $(x + \Delta x)$ is

$$Q + \Delta Q = -\alpha A\left[\frac{\partial T}{\partial x} + \frac{\partial}{\partial x}\left(\frac{\partial T}{\partial x}\right)\Delta x\right]\Delta t \tag{2}$$

Supposedly, the heat input into the sample is W per volume unit, and the loss due to irradiation and convection is neglected.

Therefore, the energy obtained in per length at Δx is

$$\Delta Q + WA\Delta x\Delta t = \alpha A\frac{\partial^2 T}{\partial x^2}\Delta x\Delta t \tag{3}$$

Consequently, the rate of temperature variation is

$$\frac{\partial T}{\partial x} = \frac{1}{\rho c}\left(W + \alpha\frac{\partial^2 T}{\partial x^2}\right) \tag{4}$$

where ρ is the density of sample and c is the specific heat capacity.

When the processing is stable, $\frac{\partial T}{\partial x} = 0$, then

$$W = -\alpha\frac{\partial^2 T}{\partial x^2} \tag{5}$$

After integration,

$$\frac{\partial T}{\partial x} = -\frac{W}{\alpha}x + b \tag{6}$$

When $x = \frac{1}{2}L$, $\frac{\partial T}{\partial x} = 0$, it obtains

$$b = \frac{WL}{2\alpha} \tag{7}$$

Integratedly, then

$$T = -\frac{W}{2\alpha}x^2 + \frac{WL}{2\alpha}x + b' \tag{8}$$

At $x = 0$ and $T = T_0$, so $b' = T_0$; therefore,

$$T = -\frac{W}{2\alpha}x^2 + \frac{WL}{2\alpha}x + T_0 \tag{9}$$

or

$$T - T_0 = \frac{W}{2\alpha}(L - x)x \tag{10}$$

At $x = \frac{1}{2}L$ and $T = T_{max}$, so

$$T_{max} - T_0 = \frac{WL^2}{8\alpha} \tag{11}$$

Assumption: $T = T_{max} - \Delta T$, then

$$T_{max} - \Delta T - T_0 = 4\frac{(T_{max} - T_0)}{L^2}(L - x)x \tag{12}$$

Therefore,

$$x = \frac{L}{2}\left[1 \pm \sqrt{\frac{\Delta T}{T_{max} - T_0}}\right] \tag{13}$$

As a result, the length between T_{max} and $T_{max} - \Delta T$ is

$$\Delta x = L\sqrt{\frac{\Delta T}{T_{max} - T_0}} \tag{14}$$

According to Eq. (14), it reveals that with the increment of temperature difference, the area between the solid phase and liquid phase increases simultaneously. However, if the temperature is too high, it will make the welding failure because the area between the solid phase and liquid phase enlarges. As a result, when the samples are impacted, the relative sliding of samples occur [21–23]. Meanwhile, the grain size and microstructure of AMCs will be too larger and coarser to decrease the property of the welded joints.

Furthermore, considering the practical situations during the welding process, the resistance between the two welded pieces due to the heterogeneous materials at the welding area is definitely higher than that of the calculation on the basis of the theory. Consequently, Eq. (14) will be

$$\Delta x = \eta L\sqrt{\frac{\Delta T}{T_{max} - T_0}} \tag{15}$$

where η is the coefficient of the influence of hetero-resistance at the welding area.

3.2. Microstructure of welded joint at various temperatures

The fractographs of SiC_p/A356 are shown in **Figure 4**. It illustrates that when the welding temperature is 563°C, the initial morphology of substrate is still obviously detected, and some sporadic welded locations appear together with some rather densely scattering bare reinforcement particles as shown in **Figure 4a**. With the welding temperature increasing to 565°C, more liquid phases form. Under the effect of pulse-impact, some wet locations in the

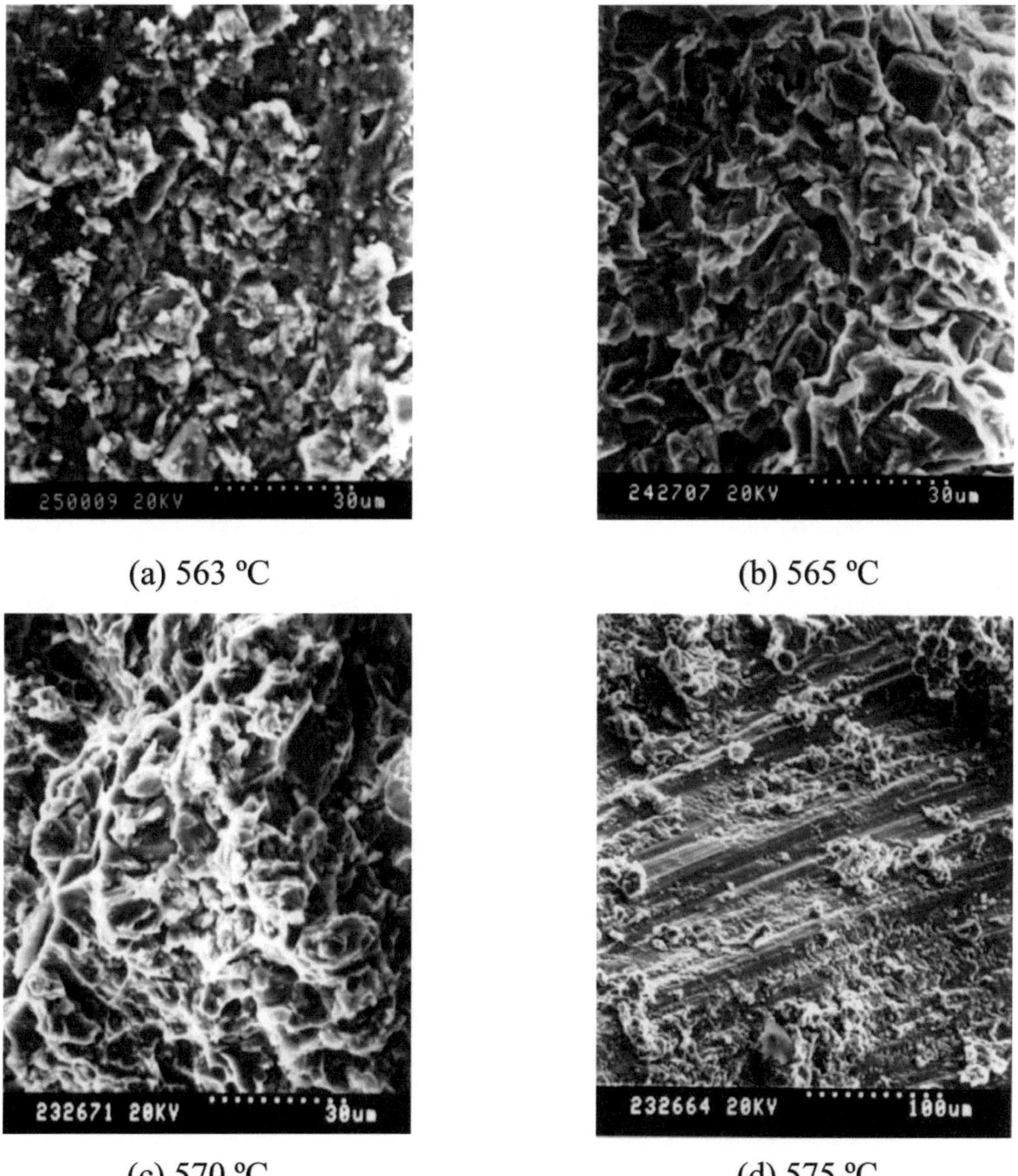

Figure 4. Fractographs of SiC_p/A356 at various temperatures. (a) 563°C, (b) 565°C, (c) 570°C and (d) 575°C.

joint excellently weld, and the condition of the aggregated solid reinforcement particles is improved. However, the bare reinforcement particles still distribute on the fractographic surface. It indicates that substrates do not weld ideally and it consequently results in a low-strength joint (**Figure 4b**).

Figure 4c shows the fractograph of welded joint at 570°C. It illustrates that the fracture is dimple fracture. Moreover, SEM image of the fracture surface shows some reinforcement particles (SiC) in the dimple. In order to confirm the state of these reinforcement particles, particles itself and matrix near to these particles were analyzed by energy-dispersive X-ray analysis (EDX). The result is expressed in **Figure 5**. It indicates that reinforcement particles (SiC) are wet by matrix alloy successfully suggesting that the reinforcement particles have been perfectly wet and the composite structure of reinforcement/reinforcement has been changed to the state of reinforcement/matrix /reinforcement.

As welding temperature increases to 575°C, it leads to more and more liquid-phase matrix alloy distributed in the welded interface. Meanwhile, more liquid-phase matrix alloy reduces the effect of impact on the interface of the welded joints; subsequently, the application of transient pulse-impact causes the relative sliding of the weldpieces that jeopardizes ultimately the formation of proper joint as shown in **Figure 4d**.

The relevant fractographies of SiC_p/6061Al and Al_2O_{3p}/6061Al at various welding temperatures are shown from **Figures 6** to **7**. It shows that the fracture surfaces under the effect of pulse-impact are similar to that of SiC_p/A356. The fractures are all dimple fractures with some reinforcement particles (SiC, Al_2O_3) in the dimple.

SEM results of the fracture surface show that the reinforcement particles have been perfectly wet and the composite structure of reinforcement/reinforcement has been changed to the state of reinforcement/matrix /reinforcement. XRD pattern of the fracture surfaces (**Figure 8**) does not show the existence of any harmful phase or brittle phase of Al_4C_3. This suggests the effective interface transfers between reinforcement particles and matrix in the welded joint that subsequently provides favorable welding strength [16–18].

3.3. Effect of pulse-impact on SiCp/6061Al LPPIDW on the interface reaction

During LPPIDW, SiC_p/6061Al welding temperature is so high that the reaction between SiC particles and aluminum matrix occurs as follows:

$$3SiC(s) + 4Al(l) \rightarrow Al_4C_3(s) + 3Si(s) \tag{16}$$

The relevant free energy [24] is

$$\Delta G\left(J \cdot mol^{-1}\right) = 113900 - 12.06TlnT + 8.92 \times 10^{-3}T^2 + 7.53 \times 10^{-4}T^{-1} + 21.5T + 3RT\ln \alpha_{[Si]} \tag{17}$$

where $\alpha_{[Si]}$ Si is the activity in the liquid of aluminum.

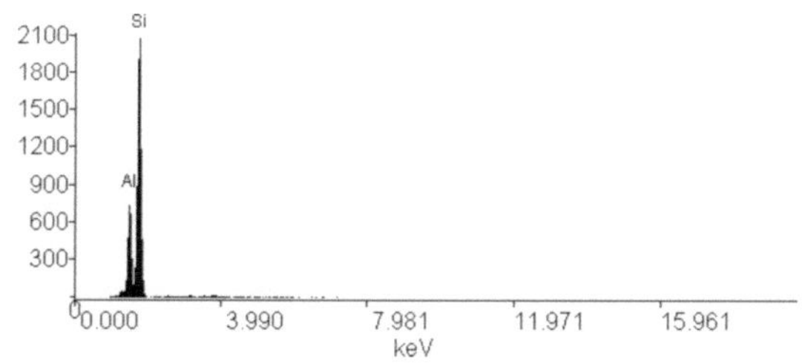

Figure 5. Energy-dispersive X-ray analysis of the fracture surface of SiC_p/A356.

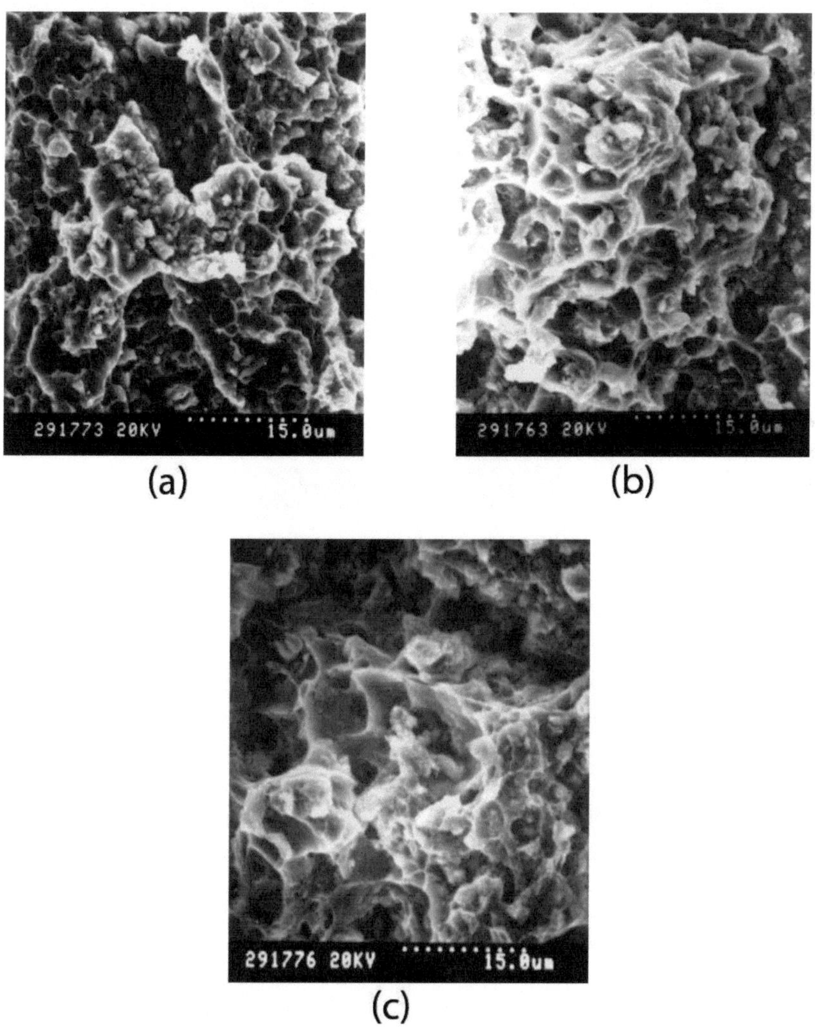

Figure 6. Fractographs of SiC_p/6061Al at various temperatures. (a) 620°C, (b) 623°C and (c) 625°C.

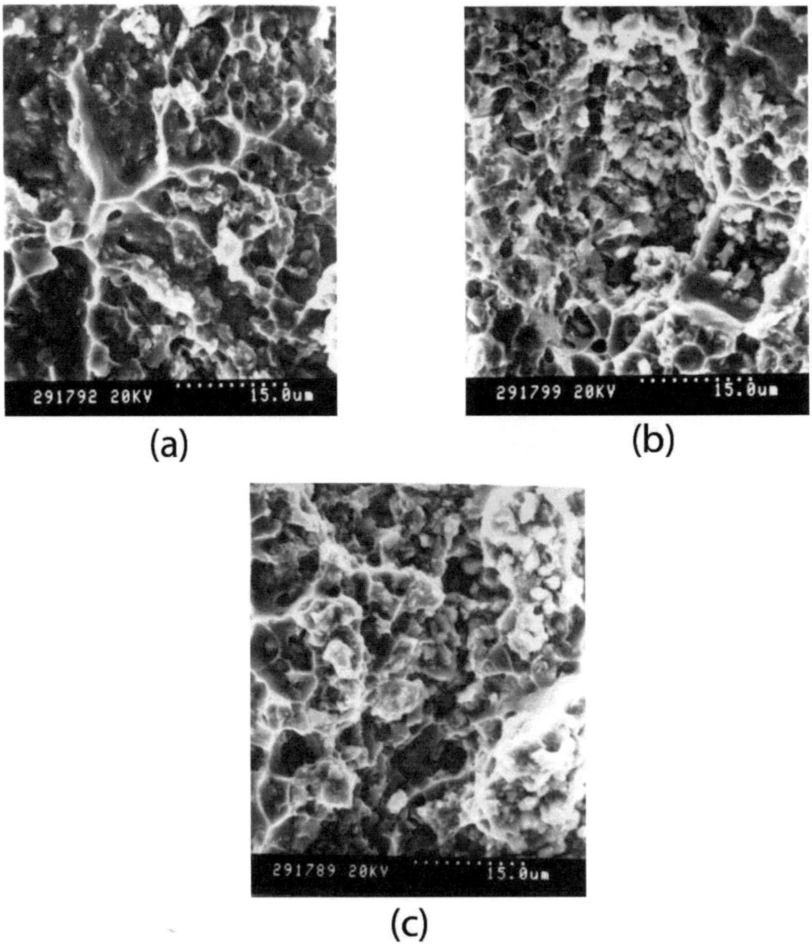

Figure 7. Fractographs of Al_2O_{3p}/6061Al at various temperatures. (a) 641°C, (b) 644°C and (c) 647°C.

In the viewpoint of thermodynamics, $\Delta G > 0$ during LPPIDW SiC_p/6061Al. Therefore, the reaction of Eq. (16) does not occur. However, in accordance with binary alloy-phase diagrams of Al-Si as shown in **Figure 9** [11], Si will dissolve in the aluminum matrix during the welding, and the activity of Si in the liquid aluminum varies at various temperatures. The relationship between free energy and welding temperature in the interface after considering Si dissolution in the aluminum matrix is shown in **Figure 10**, which indicts that Eq. (16) will occur in the welding.

It is well known that the interfaces between the reinforcement (particles) and matrix are playing the extremely vital role in AMCs, especially in adjusting their matches. The slight reaction at the interfaces is propitious to improve the property of welded joints due to the release of internal stress, which is caused by the hetero-matches between the

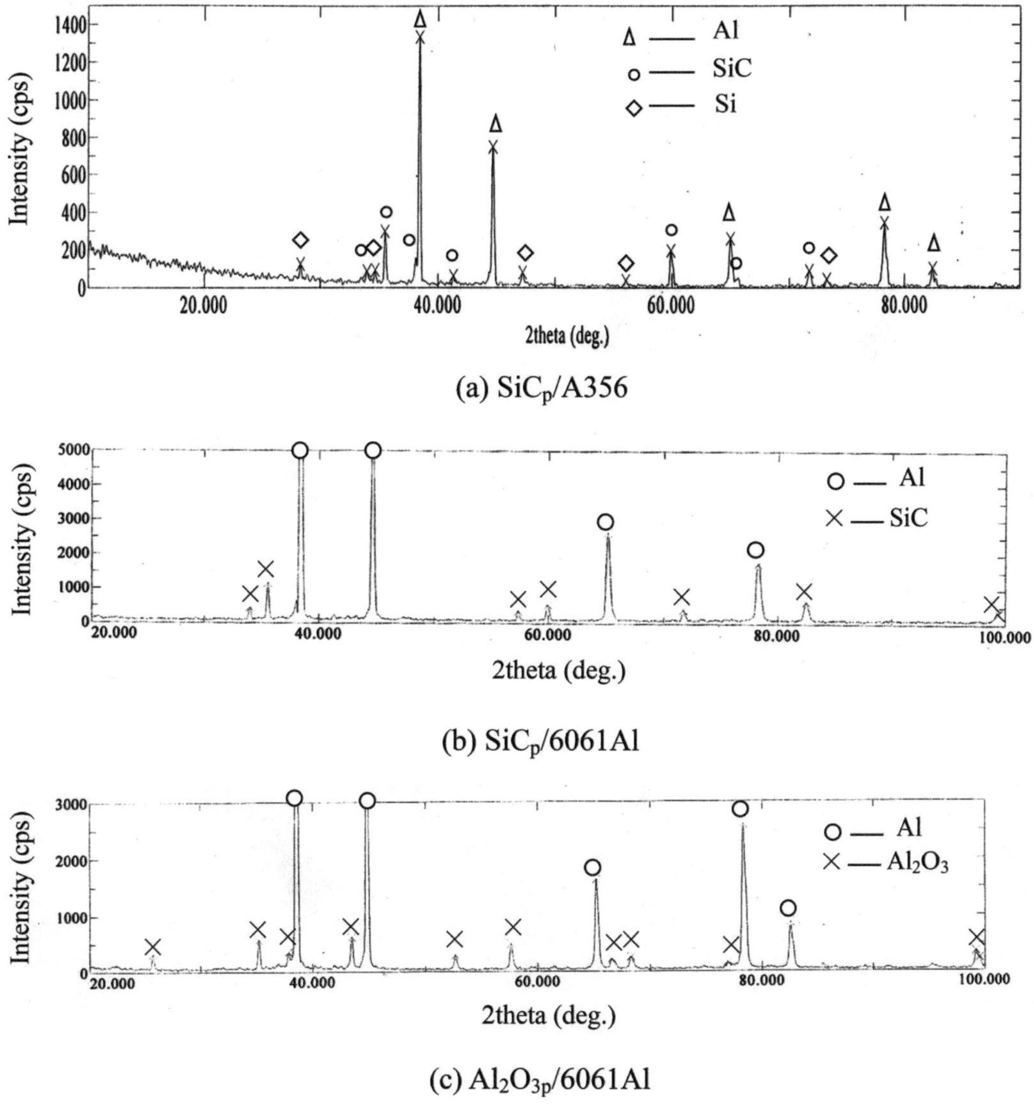

Figure 8. XRD pattern of the fracture surfaces. (a) SiC_p/A356, (b) SiC_p/6061Al and (c) Al_2O_{3p}/6061Al.

reinforcements and matrix. The distribution of dislocations at the interface is shown in **Figure 11**. It illustrates that the density of dislocations decreases remarkably, especially compared with that of its nearing area, which elucidates that the effective reaction at the interface occurs and releases the internal stress due to the hetero-matches between the reinforcement/particle (SiC) and matrix. Consequently, the property of welded joints improves compellingly and convincingly.

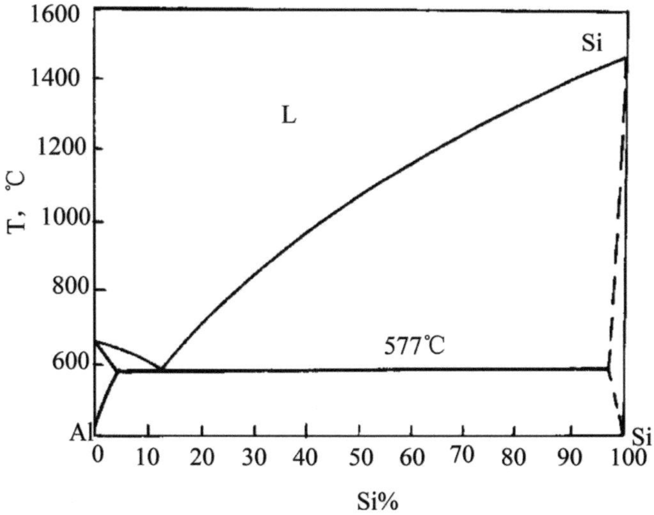

Figure 9. Binary alloy phase diagrams of al-Si [25].

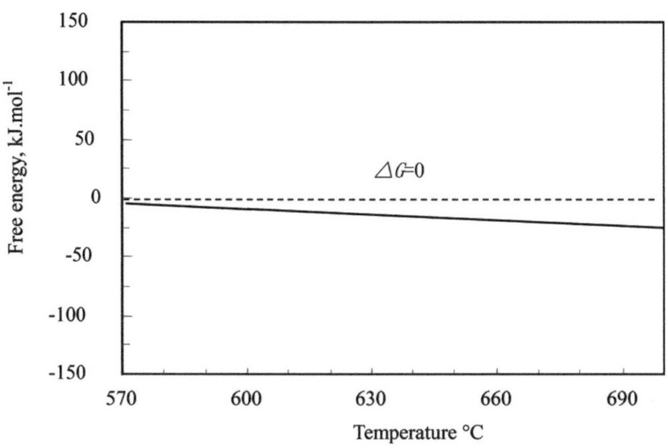

Figure 10. Relationship between free energy and welding temperature in the interface.

Also, the results shown in **Figure 6** of SiC_p/6061Al are better than that of Al_2O_{3p}/6061Al shown in **Figure 7** just due to this interfacial reaction between the reinforcement and matrix, which releases the thermal mismatch stress to an acceptable extent between the reinforcement and matrix to allow load transfer from the matrix to reinforcement successfully. As a result, it has advantageous effect of improving the strength of welded joints further [18].

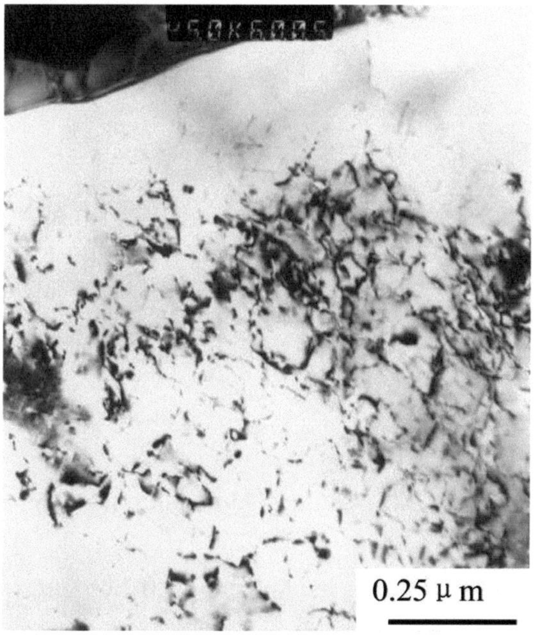

Figure 11. Distribution of dislocations at the interface.

4. Conclusion

The fractography results of liquid-phase-pulse-impact diffusion welding of particle reinforcement aluminum matrix composites (SiC_p/A356, SiC_p/6061Al, and Al_2O_{3p}/6061Al) show that:

(1) The area of the solid-phase and liquid-phase coexistence in the welding can be calculated by $\Delta x = \eta L\sqrt{\frac{\Delta T}{T_{max} - T_0}}$

(2) The fracture of welded joints with optimal processing parameters is the dimple fracture with some reinforcement particles (SiC, Al_2O_3) in the dimples.

(3) Distinctly clear interface between reinforcement particle and matrix overcomes some diffusion problems normally encountered in conventional diffusion welding and prevents the formation of harmful microstructure or brittle phase in the welded joint.

(4) The slight reaction occurs at the interfaces of SiC_p/6061Al, which is propitious to improve the property of welded joints because of the release of internal stress caused by the hetero-matches between the reinforcements and matrix.

Author details

Kelvii Wei Guo[1,2*]

*Address all correspondence to: guoweichinese@yahoo.com

1 State Key Laboratory of Millimeter Waves (Partner Laboratory in City University of Hong Kong), City University of Hong Kong, Kowloon Tong, Kowloon, Hong Kong

2 Department of Mechanical and Biomedical Engineering, City University of Hong Kong, Kowloon Tong, Kowloon, Hong Kong

References

[1] Nair SV, Tien JK, Bates RC. SiC-reinforced aluminium metal matrix composites. International Metals Reviews. 1995;**30**(6):275-288

[2] Avettand-Fènoël MN, Simar AA. Review about friction stir welding of metal matrix composites. Materials Characterization. 2016;**120**:1-17

[3] Pandey U, Purohit R, Agarwal P, Dhakad SK, Rana RS. Effect of TiC particles on the mechanical properties of aluminium alloy metal matrix composites (MMCs). Materials Today: Proceedings Part D. 2017;**4**(4):5452-5460

[4] Butt J, Mebrahtu H, Shirvani H. Microstructure and mechanical properties of dissimilar pure copper foil/1050 aluminium composites made with composite metal foil manufacturing. Journal of Materials Processing Technology. 2016;**238**:96-107

[5] Loyd DJ. Particle reinforced aluminum magnesium composites. International Materials Reviews. 1994;**39**(1):1-22

[6] Rana RS, Purohit R, Soni VK, Das S. Characterization of mechanical properties and microstructure of Aluminium alloy-SiC composites. Materials Today: Proceedings. 2015;**2**(4–5):1149-1156

[7] Pirondi A, Collini L. Analysis of crack propagation resistance of Al-Al_2O_3 particulate-reinforced composite friction stir welded butt joints. International Journal of Fatigue. 2009;**31**(1):111-121

[8] Rotundo F, Ceschini L, Morri A, Jun TS, Korsunsky AM. Mechanical and microstructural characterization of 2124Al/25vol.%SiCp joints obtained by linear friction welding (LFW). Composite Part A: Applied Science and Manufacturing. 2010;**41**(9):1028-1037

[9] Gómez de Salazar JM, Barrena MI. Dissimilar fusion welding of AA7020/MMC reinforced with Al_2O_3 particles. Microstructure and mechanical properties. Materials Science and Engineering A. 2003;**352**:162-168

[10] Bataev IA, Lazurenko DV, Tanaka S, Hokamoto K, Bataev AA, Guo Y, Jorge Jr. AM. High cooling rates and metastable phases at the interfaces of explosively welded materials. Acta Materialia 2017;**135**: 277–289.

[11] Maity J, Pal TK, Maiti R. Transient liquid phase diffusion bonding of 6061-15 wt% SiCp in argon environment. Journal of Materials Processing Technology. 2009;**209**(7): 3568-3580

[12] Schell JSU, Guilleminot J, Binetruy C, Krawczak P. Computational and experimental analysis of fusion bonding in thermoplastic composites: Influence of process parameters. Journal of Materials Processing Technology. 2009;**209**(11):5211-5219

[13] Sundaram NS, Murugan N. Tensile behavior of dissimilar friction stir welded joints of aluminum alloys. Materials & Design. 2010;**31**(9):4184-4193

[14] Arik H, Aydin M, Kurt A, Turker M. Weldability of Al_4C_3-Al composites via diffusion welding technique. Materials & Design. 2005;**26**(6):555-560

[15] American Welding Society. Welding Handbook Miami: American Welding Society; 1996.

[16] Chamanfar A, Pasang T, Ventura A, Misiolek WZ. Mechanical properties and microstructure of laser welded Ti-6Al-2Sn-4Zr-2Mo (Ti6242) titanium alloy. Materials Science and Engineering: A. 2016;**663**:213-224

[17] Wert JA. Microstructures of friction stir weld joints between an Aluminium-Base metal matrix composite and a monolithic Aluminium alloy. Scripta Materialia. 2003;**49**(6):607-612

[18] Fernandez GJ, Murr LE. Characterization of tool wear and weld optimization in the friction-stir welding of cast aluminum 359+20% SiC metal-matrix composite. Materials Characterization. 2004;**52**(1):65-75

[19] Hsu CJ, Kao PK, Ho NJ. Ultrafine-grained al–Al_2Cu composite produced in-situ by friction stir processing. Scripta Materialia. 2005;**53**(3):341-345

[20] Marzoli LM, Strombeck AV, Dos Santos JF, Gambaro C, Volpone LM. Friction stir welding of an AA6061/Al_2O_3/20p reinforced alloy. Composites Science and Technology. 2006;**66**(2):363-371

[21] Guo W, Hua M, Ho JKL. Study on liquid-phase-impact diffusion welding SiC_p/ZL101. Composites Science and Technology. 2007;**67**(6):1041-1046

[22] Guo W, Hua M, Ho JKL, Law HW. Mechanism and influence of pulse-impact on properties of liquid phase pulse-impact diffusion welded SiCp/A356. The. International Journal of Advanced Manufacturing Technology. 2009;**40**(9–10):898-906

[23] Guo W, Hua M, Law HW, Ho JKL. Liquid phase impact diffusion welding of SiC_p/6061Al and its mechanism. Materials Science and Engineering: A. 2008;**490**(1–2):427-437

[24] Isekl T, Kameda T, Maruyama T. Interfacial reaction between SiC and aluminum during joining. Journal of Materials Science. 1984;**19**:1692-1698

[25] Ohring M. Engineering Materials Science. San Diego: Academic Press; 1995